History of Physics

The Springer book series *History of Physics* publishes scholarly yet widely accessible books on all aspects of the history of physics. These cover the history and evolution of ideas and techniques, pioneers and their contributions, institutional history, as well as the interactions between physics research and society. Also included in the scope of the series are key historical works that are published or translated for the first time, or republished with annotation and analysis.

As a whole, the series helps to demonstrate the key role of physics in shaping the modern world, as well as revealing the often meandering path that led to our current understanding of physics and the cosmos. It upholds the notion expressed by Gerald Holton that "science should treasure its history, that historical scholarship should treasure science, and that the full understanding of each is deficient without the other." The series welcomes equally works by historians of science and contributions from practicing physicists.

These books are aimed primarily at researchers and students in the sciences, history of science, and science studies; but they also provide stimulating reading for philosophers, sociologists and a broader public eager to discover how physics research – and the laws of physics themselves – came to be what they are today.

All publications in the series are peer reviewed. Titles are published as both print- and eBooks. Proposals for publication should be submitted to Dr. Angela Lahee (angela.lahee@springer.com) or one of the series editors.

More information about this series at http://www.springer.com/series/16664

Chris Talbot · Olga Pattison
Editors

Boris Hessen: Physics and Philosophy in the Soviet Union, 1927–1931

Neglected Debates on Emergence and Reduction

Edited and Translated by Chris Talbot and Olga Pattison

Springer

Edited and Translated by
Chris Talbot
Abingdon, UK

Olga Pattison
Oxon, UK

ISSN 2730-7549 ISSN 2730-7557 (electronic)
History of Physics
ISBN 978-3-030-70044-7 ISBN 978-3-030-70045-4 (eBook)
https://doi.org/10.1007/978-3-030-70045-4

This Springer imprint is published by the registered company Springer Nature Switzerland AG
The registered company address is: Gewerbestrasse 11, 6330 Cham, Switzerland

Preface

These translations are intended to make available for the English reader writings of Boris Hessen on the philosophy of physics. With the exception of the last chapter, which is a conference report, they were all published in Russian in the four years before Hessen gave his paper, well known to historians of science, in 1931 at the Second International Congress of the History of Science in London, entitled "On the Social and Economic Roots of Newton's Principia."

It is a fact that whereas dozens of books and articles have referred to Hessen's London paper, some favorable and some critical, they have been focused on the impact it made on the historiography of science. The approach supposedly originated by Hessen is that science is based on two factors, the socio-economic context or external factor in addition to the internal development of the subject itself. Hessen's own context and the content of his own work in the years leading up to 1931 are, with a few notable exceptions, rarely considered.

This is, to say the least, one sided. Marxism was making an increasing impact throughout the world in the 1920s, following the Russian revolution. Its founders had always viewed their theory as scientific. Friedrich Engels tried to master the natural scientific ideas of his day, arguing that science could be given a better conceptual foundation when viewed from his and Marx's philosophical perspective. Yet the 1920s were witnessing perhaps the greatest ever expansion of scientific ideas, especially in the physics of relativity and quantum mechanics, which seemed to have no connection whatsoever with Marxism. Most of the leading scientists of the day either supported some form of positivism or eschewed philosophy entirely.

It was an anomaly that Boris Hessen and other Soviet philosophers were attempting to deal with. Could the latest ideas in physics, strange and apparently counterintuitive, be comprehended and even developed with the aid of a philosophy that might be relevant for social and economic questions but was surely out of date when it came to physics?

Hessen was uniquely placed for the task of making a Marxist interpretation of modern physics. He was of the revolutionary generation and had taken an active part in the October events. He was a member of the Elisavetgrad Soviet in 1917 and fought in the Red Army during the civil war. He owed his philosophical training to the principal Soviet Hegelian Abram Deborin. He worked through the newly translated notes of Engels's Dialectics of Nature. He studied physics at a postgraduate level under Leonid Mandelstam, one of the leading scientists of the day. By the late 1920s, Hessen, now in his 30s, was well prepared to pioneer the development of the Marxist philosophy of science.

But a drastic change was taking place in the Soviet regime as Stalin assumed complete power and began to take a personal interest in philosophy by the end of 1930. His crude instrumentalism demanded that science was subordinated to the needs of technology and that philosophy was directed by the Politburo. Hessen was forced to abandon his work on modern physics and was sent to the London conference, having been told to concentrate on promoting Marxism and on the importance of technology for scientists. Given the circumstances he gave a remarkable performance managing, by implication, to oppose those in the USSR who attacked modern physics because of the "bourgeois" philosophy of its leading proponents. Needless to say, it was not enough to spare his life and he was executed in 1936 at the age of only 43.

We are indebted to many people who have helped to make this book possible. Angela Lahee of Springer for supporting the publication of Hessen's writings and an anonymous Springer reviewer who wrote that the material was a "treasure trove of pretty solid ideas about the foundations of physics." Chris would like to thank historians of science who have expressed interest and support for this project, especially Olival Freire Junior. He would also like to thank the staff of the Radcliffe Science Library, University of Oxford, who, in the days before Covid were always helpful. Special thanks must go to librarian Gabriele Luckey Nichols for her help with translating scientific German. Both of us would like to express our indebtedness to Mrs. Tatiana D. Maslova, Deputy Director of the Moscow State University Library, who helped us to get hold of her library's archival material, and to Mr. Nikolai S. Matveyev whose expertise in making three-way translations between Russian, German, and English with difficult texts in physics was invaluable. We would also like to thank Prof. S. N. Korsakov for kindly allowing us to translate Hessen's report to the Odessa Conference of physicists in 1930, which he found in the archives, included here as Chap. 12.

The last 10 months with a global pandemic have been a demanding period in which to complete this book. Chris would like to thank his three children, their partners, and four grandchildren for their loving support. His wife Ann, as always, gave unstinting encouragement for a project which has consumed much time. Olga thanks her husband Julian for his patience, her daughter Anna for her advice, her brother Gregory for procuring the material in Russian, and her granddaughter Katya for her inspiring inquisitiveness.

Abingdon, UK Chris Talbot
Oxon, UK Olga Pattison

Contents

Note on the Translations

Translations from the Russian originals are given here in Chaps. 2– 12. We have tried to bring out the philosophy and physics of Boris Hessen's writings.

All of Hessen's references and footnotes, which tend to be quite sparse, have been reproduced here. His footnotes have been prefixed by BH or in the case of Chaps. 2 and 3, co-authored with Vasilii Egorshin, prefixed by BH and VE.

All footnotes added by us are prefixed by TN. We have managed to find references for virtually all the quotations given by Hessen and have given citations in our footnotes to the references at the end of each chapter. For quotations in French or German we have given references to English translations if these are available. We have also added comments in our footnotes to assist the reader where it seemed appropriate. Where possible we have used the original English, or published translations from the French or German, in the quotations. Where Hessen's Russian translations are poor, and English translations are not available, we have given translations from the Russian supplemented by giving the original French or German in our footnotes.

There are three key Russian words in this project that can be translated into English in more than one way. Upon reflection, we made the following choices:

- *sovokupnost' an aggregate* rather than a set or a totality.
- *sluchainost' a chance (noun)* rather than a contingency or an accident.
- *sluchainy accidental* rather than random or chance (adjective).

We have added plates of most of the participants in the Fifth Congress of Russian physicists featured in Chap. 2 at the end of that chapter. Unfortunately there appear to be no good quality photographs of Boris Hessen, either individual or in a group.

Extra comments have been added at the end of Chaps. 5, 9, and 10 to give clarification and background material.

Chris Talbot
Olga Pattison

Chapter 1
Introduction

Chris Talbot

This book arose out of a study of David Bohm's *Causality and Chance in Modern Physics*.[1] Transcribing the letters he wrote to three female friends from his exile in Brazil and then Israel—he was under threat because of his Communist sympathies in the McCarthy era—I realised that Bohm's book was essentially an attempt to relate Marxist philosophical ideas to modern physics, influenced by Engels's *Dialectics of Nature*[2]and a "materialist" reading of Hegel. Although this is clear from the letters, Bohm didn't openly admit it. The letters show he was disappointed that scientists in the Soviet Union were not impressed by his work. Bohm's success in developing an alternative "causal" approach to quantum mechanics was apparently seen there as a "classical interpretation"[3], presumably looking back to Newtonian physics and hence "mechanical". This was not Bohm's intention as a reading of *Causality and Chance* easily demonstrates.

Why was there this response in the Soviet Union? To gain some background I studied a number of reference books on Soviet science[4] and realised that, despite the impressive developments being made in physics, Soviet physicists faced enormous dangers from the 1930s through to the 1950s. Many were executed or sent to labour camps. Just as the charlatan Lysenko and his supporters dominated biology, similar ideologically motivated impostors attacked quantum mechanics as "idealist" and wanted it removed from university study. Perhaps Soviet physics survived better than

[1] See Talbot (2017). The original is Bohm (1957).
[2] *Engels* (1988).
[3] See for example Fock's views in Graham (1966).
[4] For example Graham (1971), Josephson (1991), Joravsky (2019), Bakhurst (1991).

C. Talbot (✉)
Abingdon, UK
e-mail: chris.talbot@protonmail.com

C. Talbot and O. Pattison (eds.), *Boris Hessen: Physics and Philosophy in the Soviet Union, 1927–1931*, History of Physics,
https://doi.org/10.1007/978-3-030-70045-4_1

Soviet biology, but this was because Stalin wanted a Soviet atomic bomb.[5] These were not the conditions in which Bohm's alternative Marxist approach to physics could get a hearing.

In the course of studying Paul Josephson's book I noted that Boris Hessen's[6] writings from the late 1920s were singled out.

> The brief extracts given by Paul Josephson show an interesting attempt to take a dialectical approach to standard quantum theory, based on an analysis of the relation between 'dynamic and statistical laws'. G[H]essen also tried to develop an approach to a version of the 'ether' that was not the mechanical ether of the nineteenth century, prefiguring the approach taken by Bohm. He also worked on probability theory, examining the approach of von Mises, and also on statistical mechanics, again suggesting he was thinking in ways similar to Bohm's approach above.[7]

Josephson gave a brief exposition of Hessen's relativity monograph with short quotations.[8] Sections of this monograph are translated here in Chap. 9. Similarly he gave a summary, with quotes, of Hessen's preface to a book on quantum mechanics by the German physicist Arthur Haas.[9] This is translated here in Chap. 11. Josephson describes Hessen as "a major figure of the history of Soviet physics" who was "influential in spreading the gospel of Marxist philosophy among physicists."[10]

Although Hessen is best known for the paper he delivered to the second International Congress of the History of Science in London in 1931, entitled "On the Social and Economic Roots of Newton's Principia,"[11] Josephson's was the first, and still the only book in English, to explain some of his important work before 1931 on quantum theory and relativity "which represent the first major dialectical materialist defenses of the new physics."[12]

Perhaps it should be noted that the classic text on Soviet science and philosophy by David Joravsky[13] does have some material on Hessen, and much other useful detail besides. Unfortunately Joravsky gives an unsympathetic treatment of Marxism and related Soviet science characteristic of the 1960s Cold War period when it was written. Hessen gets a rather disparaging treatment as do other Soviet scientists who took Marxism seriously. Joravsky described the writings translated here as the "Gessen manoeuvre" which he regarded as a device deployed by the Deborin group in the philosophical debates of the 1920s. They were merely intended to gain factional advantage rather than having intrinsic value.[14]

---

[5]On this see Pollock (2006).

[6]Christopher Chilvers discovered (Chilvers 2003) that in the few records available, Hessen signed his name in English with an 'H' rather than the alternative translation from the Russian of 'G'. We have used Hessen throughout the translations here.

[7]Talbot (2017, p. 81).

[8]Josephson (1991, pp. 242–245).

[9]Ibid., pp. 266–269.

[10]Ibid., p. 240.

[11]Werskey (1973), Freudenthal and McLaughlin (2009).

[12]Josephson (1991, p. 240).

[13]Joravsky (2019).

[14]Joravsky (2019, pp. 185–187, 285–286).

In the notes at the end of his book Josephson refers to a translation of Hessen's work that would be appearing in the future. As it didn't seem to be available I wrote to Josephson inquiring about it. He replied by kindly sending me an unpublished essay on Hessen, but explained that the translation would not be going ahead. Although I have continued researches on David Bohm, I decided that Hessen's writings on philosophy and physics were vital for a deeper understanding of the theoretical background to Bohm's approach. This is the case even though Bohm had no knowledge of his existence and apart from reading Engels apparently received all his knowledge of Marxism via the Communist Party. Under Stalin the philosophical debates of the 1920s were written out of history.

Having explained my own reasons for the study of Hessen's writings, I should point out also that their publication is a contribution to the history of ideas and philosophy of science in the twentieth century. Hopefully, they will be of interest to a wide audience. Convinced that they should be made available in English I was fortunate enough to gain the help of a professional translator, Olga Pattison, in putting together the material included here.

My researches on Bohm demonstrated that much material is now available on the history of Soviet science and a more objective assessment than could be made in the Cold War years is now possible. As well as those cited above there are, for example, a later work by Graham,[15] the book by Alexei Kojevnikov[16] and the book on the tragic fate of the brilliant young physicist Matvei Petrovich Bronstein.[17] The history of Soviet philosophy is more difficult. However Yehoshua Yakhot's *The Suppression of Philosophy in the USSR (The 1920s and 1930s)*[18] was translated by Frederick Choate in 2012 and is to be recommended. Yakhot, a post World War Two Soviet philosopher, tried to get the history of his subject in the 1920s and 1930s reconsidered but made no headway. Though still a convinced Marxist, Yakhot left the Soviet Union for Israel in 1975. His book was published in Russian in the US in 1981, and only appeared in Russia, much to Yakhot's surprise, in 1991 as the Soviet Union was collapsing.

Yakhot gives a thorough explanation of how the serious studies in Marxist philosophy of the 1920s, to which Hessen's work here is related, were completely undermined and transformed by the Stalinist regime in the 1930s. At first the attack was on philosophers, like Hessen, who were supporters of the leading Soviet philosopher Abram Deborin. Stalin attacked them as "Menshevising Idealists". Soon the attacks were extended to the opponents of the Deborinites, the so-called Mechanist philosophers. Stalin focused on his claim that the theoretical front was lagging behind the successes of "practical construction." Stalinist hacks, called "Bolshevizers", took over the control of philosophy and attacked the professors for not having "partisanship", i.e. subordination of their work to the demands of the central leadership. The method of criticism and self-criticism was developed, philosophers had to make

[15]Graham (1993).

[16]Kojevnikov (2004).

[17]Gorelik and Frenkel (1994).

[18]Yakhot (2012).

"confessions" concerning their alleged "errors". Philosophy itself was debased and became a weapon to police all areas of thought. In his concluding chapter, Yakhot writes,

> Under the guise of intensifying the ideological struggle, philosophers actively intervened in the various fields of scientific knowledge – genetics, physics, statistics, sociology and so forth. And everywhere this produced dramatic, and sometimes tragic consequences.[19]

Perhaps naively I had hoped to give more details and translations relating to the philosophical debates that took place in the relatively liberal period of the New Economic Policy (NEP) in the 1920s that form the context for Hessen's work. This turned out to be far too onerous a task to carry out with limited resources. It would require several books and a program of translation to do justice to these debates. What soon becomes clear from reading the Marxist approach of Yakhot, and the more detailed non-Marxist treatment by Joravsky—both have brief summaries and quotations from various philosophers—is the sheer volume and range of material involved.

The Deborin group, or Dialecticans as they were also known, to which Hessen belonged, were a theoretically cohesive group but very little even of Deborin's own work is available in English.[20] Their opponents, the Mechanists, were a much more diverse group, and each individual member would require a detailed study. It should be stressed that all these philosophers regarded themselves as Marxists. We concentrated on Hessen because he is the best known and because he was working on the Marxist philosophy of physics which relates to David Bohm. It must be stressed however that there is no intention to "take sides" in the debates, often rancorous, that took place over four or five years, or to suggest that Hessen's approach to Marxism is the only valid one. Thus we decided to omit one translation of Hessen's material[21] which was a one-sided polemic against a number of Mechanists and where we could not do justice to the various views of the opponents involved.

Both Yakhot and Joravsky show that the key question which arose in the 1920s debates was that which is now called "emergence" versus "reduction." We hope that Chaps. 6–8 give an interesting example of Hessen defending emergence, especially in relation to thermodynamics and statistical mechanics, against the Mechanist Stepanov, a committed reductionist. In current philosophy of science there has been an exponential growth of books and papers on this topic but it is disappointing that, so far as we are aware, there is no mention of Hessen and very little on the extensive Soviet debates of the 1920s.[22] As the subtitle of our book indicates, the Soviet debates of the 1920s appear to have been forgotten. The publication of Hessen's work may serve as a reminder of their interest.

---

[19] Yakhot (2012, p. 219).

[20] For an introduction to Deborin see Ahlberg (1961).

[21] "Dialectics of Nature" by Boris Hessen and Vasilii Egorshin, *Under the Banner of Marxism*, 1927, 2–3, pp. 211–225. This was directed against the 1927 edition of *Dialectics in Nature*, a collection of Mechanist writings published each year from 1926 to 1929 by the Timiryazev Institute (see Yakhot 2012, p. 27).

[22] For example in Blitz (1992), a widely used reference on the history of the topic.

Hessen's London paper "On the Social and Economic Roots of Newton's Principia," is still recognised as having had a considerable impact in the history of science in the West.[23] Yet there is a clear difference between it and the work translated here, as Loren Graham pointed out.[24] Hessen's 1931 text was "decidedly atypical of what he had been doing in the Soviet Union."[25] To gain more understanding of this, Graham advises historians of science to apply the same "externalism" towards Hessen which they discovered in Hessen's approach to Newton. What were the conditions in the Soviet Union that gave rise to the London paper? We are now in a position to add even more to Graham's explanation of key events of 1930 and 1931.

Yakhot's book gives us much more understanding in relation to philosophy. The last months of 1930 through to the London conference in July 1931 must have been a gruelling time for Hessen. All the Dialectican philosophers, including Hessen, were put under huge pressure, made to criticise their own positions and watch their leader Deborin being forced to admit his "mistakes" in the most degrading manner, from which he never recovered. They knew that this treatment was directed by Stalin and almost certainly knew about the imprisonment and torture of leading "experts" such as the economist Isaak Illich Rubin.[26]

As a physicist or a philosopher of physics Hessen faced added pressures. Here Loren Graham's explanation of the role of Ernst Kol'man, a mathematician and philosopher and leading Stalinist ideologist, must be considered. Graham interviewed Kol'man in 1971, kept in touch with him and read his articles published in 1930 and 1931 directed against Hessen. He was the Communist Party secretary responsible for the delegation who went to the London conference. It was a high level delegation which included former Soviet leader Bukharin, already in disgrace, as well as leading physicist Abram Joffe. Kol'man's job was to report back to Moscow. Before the conference Kol'man had published two articles directed against Hessen. On Hessen's writings on relativity Kol'man wrote that one must recognise that the "most harmful and dangerous of all things is empty, naked theoretization". Furthermore Stalin had made clear that "technology in the current stage decides everything." A correction of errors was necessary, "there is no Bolshevism in Hessen's science, nor in that of his comrades." After this denigration, with support from the top, it was not surprising that Hessen avoided philosophy and even modern physics in his conference contribution. What still seems amazing today is the way he dealt with Newton in socio-economic terms, then a little understood area, with very little time for preparation[27] and the courage with which he delivered his paper, knowing that his every action was being watched.

---

[23] For example in a *Reader's Guide to the History of Science* (Hessenbruch 2000) there are four entries referring to Hessen's paper.

[24] See Graham (1985).

[25] We are assuming that the Haas Preface, Chap. 11, although published in 1931 was written earlier so that the significant change in Hessen's work took place after the August 1930 Odessa conference (Chap. 12).

[26] Tucker (1990, p. 170).

[27] A small amount of background to Newton's ideas were written earlier. See Chap. 4, p. 49, n. 17 and Chap. 11, p. 145, n. 10.

Many weaknesses can no doubt be found in the paper, but Hessen's conception was at complete variance with the histories of science of the day. And as Graham points out he was able to achieve two objectives. First, he could establish his Marxist "orthodoxy" for his Soviet critics with the stress on technology. Second, by using Newton's science as an example, since unlike Einstein's it could hardly be challenged by Hessen's critics, he was able to demonstrate that science could be valid whatever the ideological and religious views of the scientist might be.

The continuity between Hessen's earlier work and the London conference paper is also an important question which Graham underlines. In his articles, Hessen "maintained that a separation could be made between the intellectual content of a theory and the social context in which it was produced." The separation which Hessen insists exists between context and content "sounds very similar to the one which the more outspoken external historians of science in the West, citing his work, would eventually question." Graham points to the article translated here in Chap. 3 in which Hessen and Egorshin[28] write that it would be "imprudent to reject the physical content" of Heisenberg's theories because he invoked "Machist" i.e. positivist philosophy.[29] In the London paper, Graham maintains that Hessen "wished to differentiate the social origins of science and its cognitive value".

As well as the connection with the "internal–external" debate a broader question should be raised. Hessen's writings are part of a tradition in the philosophy of science. They belong to a metaphysical view which may be broadly termed "organicist", usually associated with Hegel. Hessen kept close to Lenin's *Materialism and Empirio-criticism*,[30] especially Chap. 5, "The Recent Revolution in Natural Science and Philosophical Idealism," in his attempts to defend "matter" and "motion" as philosophical concepts when new physical theories of "matter-energy" were being rapidly developed and seemed to give support to positivism. He also used the new 1925 publication by Deborin of part of Lenin's *Philosophical Notebooks*[31] in which Lenin had attempted to give a materialist reading of Hegel. It is obvious from the writings here that Hessen thoroughly absorbed the material in the collection of Engels's notebooks on science, *Dialectics of Nature*,[32] a Russian version of which was also published in 1925.

Engels deserves a better treatment than he has often had by both Marxist and non-Marxist writers. In this regard the biography entitled "The Frock-Coated Communist' by Tristram Hunt is to be recommended for a sympathetic approach.[33] Hunt notes that "Engels's early years were spent in the emotional grip of the romantics, his mid-

---

[28] See Appendix on Egorshin.

[29] Chapter 3, p. 33. Graham's translation—"throwing out the physical contents of the theories"—is slightly different to ours. Note that Hessen follows Lenin in distinguishing a "philosophical concept of matter and matter as a natural scientific category", Chap. 11, p. 141.

[30] Lenin (1977).

[31] Lenin (1976).

[32] Engels (1988).

[33] Hunt (2010).

dle age was given over to science, technology and useful knowledge."[34] Hunt shows how Engels attempted to explain the latest scientific developments—conservation of energy, cellular structure, Darwinian evolution—in terms of Hegel's dialectical philosophy.[35] Engels was not a scientist—though he was in touch with top German chemists in Manchester—and as Hunt points out, despite the many interesting topics he covers in *Dialectics of Nature*, he did make one or two howlers, such as in mathematics.

Engels was tapping into a "Naturphilosophie" approach to science, developed as a reaction to Kant in Jena in the late eighteenth century, which in recent decades has been thoroughly studied by a number of academics. A number of nineteenth century scientists and mathematicians were influenced by these "Early Romantics" (Frühromantiks)—Fichte, Schelling, Hegel, and others.[36] The list includes Oersted in electromagnetic theory,[37] and the mathematicians Grassman[38] and William Rowan Hamilton[39]—the latter two made a fundamental impact on twentieth century quantum physics. Unfortunately, this is not an area which is recognised by current philosophers of science where there is, as Frederick Beiser, a leading authority in this field, put it, a "neo-Kantian embargo against Hegel's metaphysics."[40]

Hessen pursued his work in the philosophy of physics with energy and determination. He studied physics to an advanced level in an attempt to bring out the relevance of a dialectical approach. Most scientists in the Soviet Union at that time had little sympathy for Marxism but Hessen did his best to convince them. It would be a great pity if, unlike the London conference paper, his philosophical efforts in the 1920s were completely forgotten.

Revolutionary, Physicist, Philosopher

Biographical information about Boris Mikhailovich Hessen is all too brief. Existing published material is taken from archives with no informal or personal accounts. No autobiographical or biographical material appears to survive. It is hoped that more may be unearthed, especially with growing interest from Russian researchers.[41] Fascinating material was obtained by Christopher Chilvers as part of his postgraduate studies at the University of Oxford, UK.[42] Chilvers researched the papers of J. G. Crowther, science correspondent of the Manchester Guardian and British Communist Party member who was a key UK organiser of the 1931 London conference. Crowther's papers include three hand written notes that Hessen sent him when he

[34] Ibid., p. 282.

[35] Ibid., pp. 282–292.

[36] See Beiser (2003b) for an introduction.

[37] Friedman (2006).

[38] Heuser (2011).

[39] Hankins (1980).

[40] Beiser (2003a).

[41] See Korsakov et al. (2015).

[42] See Chilvers (2003, 2007).

stayed in Moscow in 1934, with Crowther's diary referring to two meetings that took place between him and Hessen. Hessen sent two letters to Crowther in 1935, which Chilvers reproduces and also a letter sent by Crowther to Hessen in May 1936, just before his arrest. Unfortunately the letters refer only to possible collaborative work and Hessen's pleasure at the interest his paper has produced, but contain nothing more of a personal nature. The following material is pieced together from the various academic studies that refer to Hessen.[43]

Boris Hessen was born into a middle-class Jewish family on 16 August 1893 in Elisavetgrad, now Kropyvnytskyi in the Ukraine. Hessen's father was the local bank manager. Hessen attended the Elisavetgrad Gymnasium school, where he became politically radicalised and also was noted for his ability in mathematics. His school friend was Igor Yevgenyevich Tamm,[44] later to become a famous physicist.

A history of pogroms and institutionalised anti-Semitism had resulted in a situation where Jews were excluded from Russian universities by the quota system established by the Tsarist regime.[45] Taking advantage of this situation Western universities enrolled groups of fee-paying Russian Jews.[46] Hessen and Tamm were part of a large group of Russian students at Edinburgh university in 1913, studying science and mathematics. One of Hessen's professors was the applied mathematician Sir Edmund Taylor Whittaker, whose pioneering history of science book[47] he refers to in Chaps. 4 and 6. We can presume that Hessen, who according to Christopher Chilvers' investigations attended few lectures and did poorly in examinations compared to Tamm, gained some inspiration from Whittaker. He may well have come across writings of James Clerk Maxwell and William Thomson (Lord Kelvin) in his Edinburgh studies some of which are cited in these writings.

With the outbreak of World War One, Hessen and Tamm were unable to return to Edinburgh from their vacation in 1914. Hessen was not conscripted due to his poor eyesight. He clearly wanted to pursue his education in maths and physics but as a Jew was rejected by the physics and maths department of Petrograd University. He was however accepted as a student of economics at Petrograd Polytechnic, and was apparently able, in 1914–1917, to attend the University as an external scholar in maths and physics. Returning to Elisavetgrad in 1917, Hessen became a founder member of the Menshevik Internationalist[48] organisation, and in September of that

---

[43]The most detailed and up to date (in terms of archival research) is in Korsakov et al. (2015, pp. 9–14, 18–21, 74, 97–98 and 162–167). Further material can be found in Joravsky (2019, pp. 185–188, 285–286, 292–295), Josephson (1991, pp. 233–246, 266–269), Chilvers (2003, pp. 422–426), Graham (1985, pp. 707–712), Freudenthal and McLaughlin (2009, pp. 253–255), and Gorelik and Frenkel (1994, pp. 44–52).

[44]See Appendix.

[45]Bushkovitch (2011, p. 274).

[46]As one writer with personal experience explains: "For decades they had headed to the West in search of education", Weizman (1949, p. 44).

[47]Whittaker (1910).

[48]The Menshevik Internationalists broke away from the Russian Social Democratic Party (Mensheviks) in May 1917 in opposition to the war. Many Menshevik-Internationalists would leave the Mensheviks and join the Bolsheviks later in 1917. Josephson (1991, p. 241) suggests that Hessen

year he was elected to the local Soviet. Like many other revolutionary young people in that period Hessen and Tamm were not just concerned to overthrow the hated Tsarist regime but desperately wanted to end the war in which the army was completely ill-equipped and millions had died. Tamm was a volunteer field medic in the war with first-hand experience of the appalling conditions. When it became clear that the leadership of the Menshevik Internationalists would not break with the Mensheviks and would not support the October revolution thus ending the war, Hessen and Tamm left and joined the Elisavetgrad Revolutionary Committee organised by the Bolsheviks. Hessen participated in the nationalisation of the very bank where his father was manager.

In March 1918 the Ukrainian People's Republic which at that time included Elisavetgrad was brought down by a coup organised by Pavlo Skoropadskyi, the so-called Hetman of Ukraine, a strong opponent of the Bolsheviks with German backing. Hessen was forced to go underground. With the assistance of his first wife E.E. Povolotskaya he moved to the town of Shpola in Kiev province. He came back after the fall of the Hetman regime in April 1919. But in May 1919 Elisavetgrad was occupied by Grigoriev's guerillas. A former general in the Red Army, Nikifor Grigoriev led a nationalist paramilitary outfit against the Bolsheviks. They carried out pogroms of the local Jewish population, killing some 3,000 Jews before the Red Army freed Elisavetgrad in June 1919. Hessen joined a detachment of the Red Army fighting Grigoriev. He was forced to destroy all his documents including papers from Edinburgh.

Hessen moved to Moscow in August 1919, acting as a political instructor for Red Army troops. We have no information on Hessen's experiences in the Red Army but this appointment suggests he played an important role. He remained in Moscow lecturing in political economy at the Sverdlov Communist University, established to train Communist Party cadres, until 1924. Presumably he was teaching economics because of attending the course at Petrograd Polytechnic. According to Sheila Fitzpatrick,[49] there was an acute shortage of Communist teachers in the social sciences and so the same people were called on to teach occasional lectures in a number of institutions, including the Sverdlov University. As a result of the problems of teaching Marxism, by 1921 it was decided that social science schools should become training institutions for Soviet government personnel, concentrating on "technical" subjects like economics and law.[50] We presume that Hessen was involved in this kind of teaching. We have no further information on Hessen in this period, and no knowledge of what role he was playing in the Bolshevik Party or in state administration.

---

was associated with Trotsky's organisation in 1917 which would undoubtedly have led to his execution in 1936. This does not seem to be the case—by 1917 Trotsky was a member of the Mezhraionka (members were called Mezhraiontsy) or Inter-District Organisation, a group formed in 1913 which took a middle course between Mensheviks and Bolsheviks. This was a different group to the Menshevik Internationalists. The Mezhraionka merged with the Bolsheviks in late July/early August 1917.

[49] Fitzpatrick (1979, p. 68).

[50] Ibid., p. 71.

Obviously the government were concerned about non-Marxist academics, many of whom had opposed the revolution and were potentially a source of opposition. In 1922 some 160 professors were deported, allegedly because of concerns that they were conspiring against the Bolshevik regime.[51] Fitzpatrick suggests it was an exemplary action "designed to intimidate the group".[52] Throughout the New Economic Policy (NEP) period of the 1920s however, non-Marxist professors still continued to dominate the universities. This was the case in social sciences and in science and technology where there was considerable expansion. Josephson notes that before 1917 there were no more than one hundred physicists in the whole of Russia.[53] By 1930 in the Leningrad Physico-Technical Institute (LFTI) alone, which Josephson mainly writes about, the total personnel numbered 256.[54] Note that out of this total there were only 7 Party members, reflecting the situation in science as a whole throughout the Soviet Union at that time. In this period government policy was to leave scientists to develop their work without interference. For example Lunacharsky, the Commissar of Education, assured scientists in 1928 that it was their "legitimate right" not to be Marxists.[55]

Boris Hessen saw physics, rather than applied economics, as his real vocation. Recognising his limited education—it was seven years since he had studied mathematics and he had "no systematic knowledge in physics" as well as lecturing at the Sverdlov Communist University, in 1924 he began to study physics at the Institute of Red Professors (IKP). He specialized in "the structure of matter, the physics of heat and electricity, and theoretical physics". According to Korsakov he studied under the well-known Russian physicist Leonid Mandelstam,[56] and "participated in a seminar of Academician Mandelstam where he gave a paper on the law of large numbers in connection with the works of von Mises"[57] which was later published in the journal *Uspekhi Fizicheskikh Nauk* (*Advances in Physical Sciences*). Hessen also studied under Abram Joffe[58] the head of LFTI, featured here in Chap. 2, who was central to the development of a new generation of Soviet physicists in the 1920s and 30s.

We have no information on Hessen's development as a Marxist. Some knowledge of the classics and an introduction to Marx and Engels's materialist philosophy would have been obligatory for Party members. But Hessen was unusual in wanting to study further, joining the group around the philosopher Abram Deborin.[59]

---

[51]This incident occurred in 1922 when philosophers and intellectuals who were not Marxists, and their families, were exiled in the "philosophers' steamboat". Others were to follow. It was feared they could incite support for the White armies, backed by the Western powers, that the Bolsheviks had just defeated. This action was widely criticized in the West.

[52]Ibid., p. 76.

[53]Josephson (1991, p. 4).

[54]Ibid., p. 341.

[55]Joravsky (2019, p. 65).

[56]See Appendix.

[57]See Appendix.

[58]See Appendix.

[59]No date for this is known.

By 1926 Hessen embarked on a career in the philosophy of physics, becoming an assistant professor in the Department of the History and Philosophy of Natural Science of Moscow University. The department had been set up in 1923 and may well have been the first university department in the world to cover both philosophy and history of science. Specialised departments only developed after World War Two in Western countries. Unfortunately the founders of the department A. A. Maksimov and A. K. Timiryazev did not, like Hessen, embrace the new developments in twentieth century physics but rather did their best in different ways to obstruct and derail physics in the Soviet Union over the next several decades.[60] Hessen's most distinguishing feature is that he wanted to be a philosopher with a serious grounding in modern physics, so he continued his studies at the IKP until 1928. Hessen remained at the IKP as a lecturer and as deputy director of the section for natural sciences, giving lectures along with Maksimov and Egorshin. In 1927 Hessen led an initiative by the Deborin group to persuade natural scientists to take an interest in Marxist philosophy writing, at first jointly with Egorshin, articles which are presented here. In March 1928 upon finishing his studies at IKP Hessen applied to attend summer courses in Berlin, between May and September. Though Einstein, Planck and von Mises are mentioned in the application it is not clear who in fact gave these courses. Permission was granted for Hessen to attend.

In 1928 the Institute of Philosophy and the Philosophical Section of the Communist Academy were united with Deborin at their head. Hessen was appointed professor, though apparently continuing in his post as a junior professor at Moscow University. In the new Institute Hessen was promoted above Maksimov, with the latter beginning to criticise the Deborinites' scientific philosophy, claiming that modern physics itself was "idealist" and "bourgeois".

Hessen gave a report at the All-Union Congress of Physicists held in Odessa, August, 1930 (see Chap. 12). In 1930 he also travelled to the Sixth Congress of German Physicists in Konigsberg. As we saw above Hessen and the Deborinites as a whole came under increasing political attack later that year.

In 1931, the same year Hessen made the trip to London, he was appointed full Professor of Physics at Moscow University. The Institute of the History of Science and Technology was established in 1932 and Hessen became Dean of the Physics Faculty. He was elected a corresponding member of the Academy of Sciences in 1933, and from 1934 to 1936 was deputy director at the Physics Institute of the Academy of Sciences, under Sergei Vavilov.[61]

Korsakov states[62] that after 1931 "he [Hessen] fell silent as a scientist. He practically stopped publishing and speaking at philosophical meetings. He either did not finish or did not publish the final parts of the already published essays 'Mechanical Materialism and Modern Physics' and 'The Statistical Method in Physics and a New Approach to von Mises's Probability Theory'."[63] Russian versions of the London

---

[60] See Appendix on Maksimov and Timiryazev.

[61] See Appendix.

[62] Korsakov et al. (2015, p. 97).

[63] See Chaps. 6–8 and 10 translated here for the parts of these that were already published.

paper were produced in 1933 and 1934 with only minor differences from the original.[64] Although Hessen was preparing a "much enlarged and revised" version of the London paper with a view to publishing a book in English, possibly with Kegan Paul, according to Chilvers[65] it was never completed. The manuscript has apparently been lost.

It should be noted that although he did not publish, Hessen was not prepared to accept the attacks on physics that were being made by Maksimov and others. Josephson explains how, in 1934, Hessen, Joffe, and other physicists were attacked in philosophical journals. At a special session of the Communist Academy in 1934, Joffe courageously defended the physicists against the attacks, utilising Hessen's work in order to demonstrate that relativity and quantum mechanics could be interpreted in terms of dialectical materialism, and showing that the anachronistic views of Timiryazev, along with the philosophical attacks of Maksimov and others, were doing serious damage to Soviet physics.[66]

Hessen appeared to have thought that if he abandoned his philosophical publications he would be allowed to continue to play a role in the teaching and administration of science in the Soviet Union. Tragically that was not the case. He was arrested on August 21 1936, two days after the Zinoviev-Kamenev show trial began. Josephson has interviewed a physicist who was a student at the time.[67] Hessen's arrest "came as a shock to many students and faculty; Hessen was widely respected, made a good impression with students, and knew physics well enough." At a general meeting of the scientific institute of the university, called without prior notification to senior physicists or administration in August or September, young Communists and Komsomol members announced that Hessen had been arrested as a wrecker and "enemy of the people," asking all participants to contribute to the general abuse. They accused Hessen of "a lot of nonsense," said he "cultivated bad instincts among the young workers" attending the university since they drank vodka all night and claimed he was a "wrecker" of the educational program, with few qualified students graduating.

Chilvers explains that Hessen was interrogated and tortured seventeen times then brought before a secret military tribunal, chaired by the notorious V. V. Ulrikh, on December 20 1936. According to Korsakov[68] Hessen admitted participation in counter-revolutionary Trotskyist activity and his animosity toward Comrade Stalin on August 23 1936. But he denied creating a terrorist cell at Moscow State University throughout the interrogations. He was sentenced to death for membership of a counter revolutionary Trotskyist-Zinovievist terrorist organization, the standard charge against all opponents of the regime for supposedly assassinating Kirov and attempting assassination of other Party leaders. He was shot on the same day.

---

[64] See Freudenthal and McLaughlin (2009), p. 2.

[65] Chilvers (2003, pp. 432–3).

[66] Josephson (1991, p. 270).

[67] See Josephson (1991, p. 310, n. 89). Josephson interviewed Immanuil Lazarevich Fabelinskii on November 1, 1989. He graduated from Moscow University in 1936, a specialist in optics and a corresponding member of the Academy of Sciences since 1979.

[68] Korsakov et al. (2015, p. 167).

Hessen's widow Anna Ivanovna Yakovleva was left to bring up their son Andrey, born in 1929, and a daughter Eva born in 1923. His widow was expelled from the Communist Party in March 1937 and sacked from the Lenin Museum where she worked. Then she was exiled to Archangel as a family member of the enemy of the people, working there as a seamstress. In Archangel she was arrested on December 10 1937 and condemned by Troika on February 9 1938 for 8 years. She was freed in 1943 before the end of her 8 year sentence as an invalid and lived in Zagorsk, then Maloyaroslavets, in 1950 exiled to Kazakhstan, then in 1952 returned to Maloyaroslavets. She returned to Moscow in 1954 after the death of Stalin when her sentence was lifted. On October 14 1954 she applied to the General Prosecutor of the Soviet Union and requested the rehabilitation of her husband. She spent much time and energy collecting references, statements, etc. Tamm was one of those whose statement helped to clear Hessen's name. Finally Hessen had his name cleared on April 21 1956, and was reinstated posthumously as a corresponding member of the Academy of Sciences in 1957.[69]

References

Ahlberg, R. (1961). Abram Deborin: The forgotten philosopher. In L. Labedz (Ed.), *Revisionism, essays on the history of Marxist ideas, Chapter IX* (pp. 126–141). New York: Frederick A. Praeger.

Bakhurst, D. (1991). *Consciousness and revolution in Soviet philosophy, from the Bolsheviks to Evald Ilyenkov*. Cambridge: Cambridge University Press.

Beiser, F. (2003a). Hegel and naturphilosophie. *Studies in History and Philosophy of Science*, *34*, 135–147.

Beiser, F. (2003b). *The romantic imperative*. Cambridge, Massachusetts: Harvard University Press.

Blitz, D. (1992). *Emergent evolution: Qualitative novelty and the levels of reality*. Dordrecht: Springer.

Bohm, D. (1957). *Causality and chance in modern physics* (1st ed., p. 1984). Abingdon: Routledge and Kegan Paul. Second edition with new preface by Bohm.

Bushkovitch, P. (2011). *A concise history of Russia*. Cambridge: Cambridge University Press.

Chilvers, C. A. J. (2003). The dilemmas of seditious men: The Crowther Hessen correspondence in the 1930s. *The British Journal for the History of Science*, *36*(4), 417–435.

Chilvers, C. A. J. (2007). The tragedy of comrade Hessen: Biography as historical discourse. In T. Söderqvist (Ed.), *The history and poetics of scientific biography* (pp. 105–120). Aldershot: Ashgate.

Engels, F. (1988). *Dialectics of nature, in Marx Engels collected works* (Vol. 25). Moscow: Progress Publishers.

Fitzpatrick, S. (1979). *Education and social mobility in the Soviet Union 1921–1934*. Cambridge: Cambridge University Press.

Freudenthal, G., & McLaughlin, P. (Eds.). (2009). *The social and economic roots of the scientific revolution: Texts by Boris Hessen and Henryk Grossmann*. Boston studies in the philosophy of science(Vol. 278). Berlin: Springer.

Friedman, M. (2006). Kant—Naturphilosophie—Electromagnetism. In M. Friedman & A. Nordmann (Eds.), *The Kantian legacy in nineteenth century science* (pp. 51–80). Cambridge, Massachusetts: The MIT Press.

[69] Information from Korsakov et al. (2015, pp. 169–174).

Gorelik, G. E., & Frenkel, V. Y. (1994). *Matvei Petrovich Bronstein and Soviet theoretical physics in the thirties*. Basel: Springer.
Graham, L. R. (1966). Quantum mechanics and dialectical materialism. *Slavic Review, 25*(3), 381–410.
Graham, L. R. (1971). *Science and philosophy in the Soviet Union*. London: Allen Lane.
Graham, L. R. (1985). The socio-political roots of Boris Hessen: Soviet Marxism and the history of science. *Social Studies of Science, 15*(4), 705–722.
Graham, L. R. (1993). *Science and philosophy in the Soviet Union*. Cambridge: Cambridge University Press.
Hankins, T. L. (1980). *Sir William Rowan Hamilton*. Baltimore: John Hopkins University Press.
Hessenbruch, A. (Ed.). (2000). *Reader's guide to the history of science*. Chicago: Fitzroy Dearborn.
Heuser, M.-L. (2011). The significance of Naturphilosophie for Justus and Hermann Grassmann. In H.-J. Petsche, A. C. Lewis, J. Liesen, & S. Russ (Eds.), *Hermann Grassmann* (pp. 49–60). Basel: Springer.
Hunt, T. (2010). *The frock-coated communist*. London: Penguin.
Joravsky, D. (2019). *Soviet Marxism and natural science, 1917–1932* (2nd ed.). Abingdon, UK: Routledge.
Josephson, P. R. (1991). *Physics and politics in revolutionary Russia*. Berkeley: University of California Press.
Kojevnikov, A. B. (2004). *Stalin's great science*. London: Imperial College Press.
Korsakov, S., Kozenko, A., & Gracheva, G. (2015). *Boris Mikhailovich Gessen (1893–1936)*. Moscow: Nauka.
Lenin, V. (1976). *Collected works* (Vol. 38). Moscow: Progress Publishers.
Lenin, V. (1977). *Collected works* (Vol. 14). Moscow: Progress Publishers.
Pollock, E. (2006). *Stalin and the Soviet science wars*. Princeton and Oxford: Princeton University Press.
Talbot, C. (Ed.). (2017). *David Bohm: Causality and chance, letters to three women*. Berlin Heidelberg: Springer-Verlag.
Tucker, R. C. (1990). *Stalin in power: The revolution from above, 1928–1941*. New York and London: W.W. Norton and Company.
Weizman, C. (1949). *Trial and error*. London: Hamish Hamilton.
Werskey, G. (1973). *Science at the cross roads* (Reissued by Routledge, 2013). London: Frank Cass. Reissued by Routledge.
Whittaker, E. (1910). *A history of the theories of aether and electricity*. London: Longmans, Green and Co.
Yakhot, Y. (2012). *The suppression of philosophy in the USSR (The 1920s & 1930s)* (F. Choate, Trans. Russian original 1981). Oak Park, Michigan: Mehring Books.

Chapter 2
The Fifth Congress of Russian Physicists

Boris Hessen and Vasilii Egorshin

The First Congress of Russian physicists took place in Leningrad (called Petrograd at the time) in 1920. It laid the foundation for the subsequent congresses of physicists that took place almost every year. Before the war all-Russian congresses of natural scientists (and medical doctors) followed a German model; and each of them became an important event in the history of Russian science.

The 1860s when the first congress took place can be considered the most important landmark in this history.

Individual talks would be usually given at these congresses; some of them would present the research results in a certain *specialised area.* They would be followed by general talks summing up all scientific work in a certain area of science during a certain period.

In the former individual and specific issues would be discussed in working groups; in the latter some issues closely connected to the theory of cognition, the methodology or philosophy of the given area would be discussed or heard at plenary meetings.

While the Fifth Congress of physicists, being in fact an all-Union event, in its format followed this tradition, its content was less significant than that of the pre-war congresses, particularly in the *general* reports related to the scientific methodology.

We note this both because our readers would be most interested in these issues and because modern physics rests on methodology and philosophy at every step of the way.

Unfortunately, the Congress seemed to avoid these fundamental methodological issues (unconsciously in our opinion). This might be partly explained by the

TN: Translated from *Pod Znamenem Marksizma* (*Under the Banner of Marxism*), 1927, No. 1. Cited by Joravsky (2019, p. 399).
TN: On Egorshin see Appendix.

B. Hessen (✉) · V. Egorshin
Abingdon, UK
e-mail: chris.talbot@protonmail.com

C. Talbot and O. Pattison (eds.), *Boris Hessen: Physics and Philosophy in the Soviet Union, 1927–1931*, History of Physics,
https://doi.org/10.1007/978-3-030-70045-4_2

misconstrued idea that in the Soviet Union no issues are of special interest unless they have a direct technical or narrowly specialised significance.

But it is in the Soviet Union that the *new* scientific methodology (the materialistic dialectics) is developed and one should expect scientific congresses to have a more active approach to methodology. Moreover, we dare to wish our scientific congresses take on the task for wide circles of scientists to study the most important problems of Marxist methodology that has turned to the development of exact natural sciences.

Let us have a look at the relevant results of the recently finished Fifth Congress of physicists.

These three talks were given at the general meetings of the congress:

"Electrical qualities of dielectrics" by academician A. F. Joffe[1];

"Modern problems of geophysics" by academician P. P. Lazarev[2] and

"Evolution of the concept of energy" by professor V. K. Lebedinsky.[3]

The first subject is of much interest and great significance to electrical engineering. Here the work of the academician A. F. Joffe and his associates promises solutions to many yet unsolved theoretical and technical problems (dielectrics, i.e. insulators, and their electrical qualities). The second talk was dedicated to the new methods of studying geophysics based on the artificial simulation of some geophysical phenomena in laboratory conditions: sea currents, aurora etc. The third talk on evolution of the concept of energy was expected to follow a methodological approach because it was looking at the main problem of the relationship between energy and matter. The short talk by professor V. K. Lebedinsky was dedicated to the question of the change over historical periods of the "dualism" and the "monism" of energy and matter and vice versa.

The development of physics coincides with the more and more pronounced dualism of matter and energy. Energy, first in its potential form, and then also in the radiative one, is more and more "separated" from matter. Energy in radiation processes seems to acquire full independence. The theory of light quanta seems to complete the full separation of energy from matter. However, simultaneously with the separation from matter energy seems to materialise because the quanta in the theory of light quanta are considered as small material bodies; accordingly, the completed dualism leads us back to monism in the form of the materialised energy. Professor Lebedinsky[4] noted that H. Hertz supported the monistic idea of inherence of energy in matter and regarded potential energy as a form of kinetic energy.

---

[1]TN: On Joffe see Appendix.

[2]TN: For biographical notes on scientists referred to here see Appendix.

[3]see Appendix.

[4]TN: Lebedinsky is referring to the famous physicist (the first to prove the existence of radio waves) Heinrich Hertz (1857–1894). Lenin in *Materialism and Empirio-criticism*, Lenin (1977, p. 284) is pleased to note that Hertz, unlike many physicists of that time, does not agree with basing his exposition of mechanics on the theory of energy. See Hertz (1956, pp. 14–24). According to Lenin this demonstrates that it never occurs to Hertz to take a non-materialist "energetics" approach unlike the so-called Empirio-critics that Lenin is opposing. It is perhaps stretching things to claim, with Lebedinsky, that Hertz has a "monistic idea of inherence of energy in matter", but it is true that he thought that potential energy from a "different standpoint" could be conceived as kinetic energy—see Hertz (1956, p. 227).

The process of separation of energy and matter that ultimately leads to the "materialisation" of energy seemed to puzzle the speaker who finishes his talk with the call to decide once and for all whether matter is energy or energy is matter.

This question was posed by Marxism (Engels) more than half a century ago and has been successfully resolved. Lenin in his book *Materialism and Empirio-Criticism* looked closely at the crisis in modern physics and brilliantly formulated this problem. Energy is a quality (manifestation) of moving matter.[5] In his talk professor V. K. Lebedinsky does not make a distinction between a purely physical and a philosophical concept of matter. And this leads him to an erroneously posed question.

Two general meetings were dedicated to the so-called questions for discussion on "new quantum mechanics". These were the most interesting moments of the congress. We shall look at them in more detail.

Quite recently Bohr's quantum theory[6] was opposed to classical physics and electrodynamics (Newton-Maxwell), and justly so, as this theory introduced completely new principles into classical physics; this theory up until now has not organically merged with the latter. Now both this theory and the old physics as a whole, to a certain extent can be put next to each other. Bohr's theory can be called "classical" quantum mechanics because physics is not satisfied any more with combining classical physics with Bohr's theory; so "new quantum mechanics" is being developed in physics.

Professor Tartakovsky gave the first talk at the congress on the crisis in classical quantum theory. Quantum theory developed by Bohr in 1913 was extremely productive as it gave a rational explanation to the previously inexplicable empirical facts (spectral series etc.) Its victorious advancement was connected to the atomic model of Rutherford-Bohr. Thanks to its success the periodic system of elements received a rational explanation.

The fact that the quantum theory does not organically correspond to and does not agree with classical mechanics and electrodynamics and does not provide any explanation of a whole range of most important problems (e.g. physical optics[7]) was in the eyes of physicists redeemed by its internal harmony and its agreement with experiments.

As to the difficulties inexplicable by either classical mechanics or quantum theory, physicists hoped in due course to resolve them one way or another. Indeed, a whole

---

[5]TN: *Materialism and Empirio-criticism*, Lenin (1977, pp. 270–273). Lenin notes that the great German chemist Wilhelm Ostwald (1853–1932) sometimes created philosophical confusion by using the term "energy" to avoid taking a materialist standpoint. Lenin insisted that "energy …means material motion" (p. 272). The phrase "quality of moving matter" is an addition of the authors.

[6]TN: Bohr's quantum theory, meaning the attempt led by Bohr but including Arnold Sommerfeld and others between 1913 and 1925 to modify classical physics to explain the many new experimental results. For a popular account see Pais (1991), especially Chap. 10.

[7]TN: The quantum theory of optics, or more generally quantum electrodynamics (QED), was only just beginning at the time of this conference. Its formative development (1927–1930) was due to Werner Heisenberg, Pascual Jordan, Wolfgang Pauli, Paul Dirac, and others. For a popular introduction see Feynman (1983).

range of these difficulties have been clarified by the further development of Bohr's theory. But a vast number of problems remain unresolved (e.g. intensity of spectral lines, the effect of crossed electric and magnetic fields, and finally the spectra of all heavy elements, the interpretation of multiplets and their splitting in a magnetic field).[8] Moreover, the theory was expected to provide not only a qualitative but also a quantitative match with experiment; and while Bohr's quantum theory has not provided a single exact quantitative match, it has provided unprecedented qualitative explanations in the area of radiation and made perfectly justified predictions in many other areas.

Along with this, all the logical inconsistency of classical quantum theory became obvious. On the one hand, this theory contradicted conventional mechanics and electrodynamics and at the same time attempted to get along with both as much as possible (principle of correspondence); on the other hand, when solving a problem, classical quantum theory used conventional methods of the old physics and only in the end imposed the so called quantum limitations (conditions) of great arbitrariness which contradicted the original initial conditions of the old physics. This marked classical quantum theory with eclecticism and caused a just aspiration for new more orderly and integrated theories.

Generally speaking two approaches were possible: to improve and further *complicate* the atomic model within the framework of old quantum theory in order to achieve a better match with experiments, e.g. to accept that an electron rotates inside the atom, that it has its own magnetic moment etc., thus providing an explanation for some empirical facts; or to give up old quantum theory and attempt to build a "new quantum mechanics" based on fundamentally new principles in order to have a more orderly theory.

Among such attempts two new theories attracted the attention of the congress: the Heisenberg theory and the Schrödinger theory, not yet two years old.

We must look closer at these theories as they raise methodological and epistemological questions. Such questions have to be raised by the revolution in quantum theory that we briefly outlined here. Besides, as Lenin brilliantly noted at the time, it is here that physics in agonising pains gives birth to dialectical materialism.[9] Physics is not satisfied with aging mechanical materialism, it approaches dialectics but often falls into idealism. Knowledge of dialectical theory would have helped physicists to avoid these extremes. As demonstrated in the debates, to the credit of Soviet physicists we must say that they really were against the purely phenomenalistic view point which is partly inherent in Heisenberg. At the same time, here the lack of knowledge of dialectics became more obvious than anywhere else. However, sometimes the dialectical formulations spontaneously "broke through".

---

[8]TN: The "vast number of problems" with Bohr's theory are given in technical detail in Mehra and Rechenberg (1982).

[9]TN: In *Materialism and Empirio-criticism*, Lenin (1977, p. 313), Lenin wrote: "Modern physics is in travail; it is giving birth to dialectical materialism. The process of child-birth is painful".

At the last (89th) congress of German natural scientists and medical doctors Heisenberg raised the fundamental questions of his theory more sharply than it was done at our congress.[10]

Heisenberg refused to speak about the very reality of atoms and electrons, and consequently about models. He wished to "deal only with quantities directly observable in experiments". He said:

"The programme of new quantum mechanics meant to liberate it from any visual pictures (anschauliche Bilder) and to replace laws of classical kinematics and mechanics with simple relations between experimentally obtained quantities". The concept of the "position of an electron" was abandoned by new quantum mechanics and replaced by an aggregate of physically determined quantities. The electron coordinate is replaced by a two-dimensional table of possible radiation values. But in old quantum theory we dealt with the electron coordinates corresponding to the optical values. What are these coordinates replaced with here? They are replaced with a formal mathematical symbol—a two-dimensional table, a matrix with an infinite number of terms corresponding to the different states of an atom. The matrix is a symbolic image of the whole 'life history' of an atom; mathematical operations on matrices allow us to solve specific physical problems. This is the main premise for Heisenberg. We can directly observe primary (at the present time) particles of matter in cathode rays, in radioactive processes. In these processes we ascribe the same reality to electrons and protons and to all objects of the external world. Heisenberg says, "This view has proved to be false (falsch) in the course of time, which was in no way surprising, if we consider the fundamental inaccessibility to perception (prinzipielle Unanschaulichkeit) of these discontinuous elements. Electrons and atoms do *not*[11] have the same degree of immediate reality as the objects of our daily experience. The objective of atomic physics and therefore, quantum mechanics, is to research the character of physical reality peculiar to electrons and atoms".[12] This is not just about the necessity of using models but also about the reality of atoms and consistency of the atomistic picture of the world. These questions can be solved by means of physics—in itself, this is a great achievement; however, one should not forget that methodological formulation and solution of these questions are inconceivable within the framework of theoretical and experimental physics but require the involvement of both the newest achievements of physics and philosophy, i.e. dialectical materialism.

New "rational quantum mechanics", as Bohr called it,[13] provides quantitative coincidences with experiments better than the previous theory and explains a whole

[10]BH and VE: Heisenberg's speech was published in *Naturwissenschaft* No. 45. See also Bohr's speech at the Congress of Scandinavian mathematicians in August 1925. *Naturwissenschaft*, 1926, no.1. TN: i.e. Heisenberg (1926) and Bohr (1926).

[11]BH and VE: "Not" is in italics in the original.

[12]BH and VE Heisenberg (1926, p. 990).

[13]TN: Bohr uses the term "rational quantum mechanics" in the last section of the above reference in relation to the new theory he attributes to Heisenberg. This is probably what the authors are referring to but Bohr apparently regarded the latest version of the quantum theory at each stage, even prior to 1925, as attempting a "rational" generalisation of classical mechanics. See Bokulich and Bokulich (2005).

range of facts (e.g. half-quantum numbers) that were inexplicable in classical quantum theory.

It is easy to see the dialectical aspects in its physical content; similarly, it is obvious that in methodological issues Heisenberg arrives at pure phenomenology and limitation of physics by a formal mathematical description, i.e. ultimately at Machism and idealism.

As we mentioned earlier no one at our congress raised these questions very sharply. Most speakers in the debates and the speaker himself (I. E. Tamm) were inclined to consider Heisenberg's theory as a first step that did not exclude other methods using new types of models etc. Most participants chose not to share the "phenomenalistic" and "agnostic" approach favoured by the Heisenberg theory. But we are sorry to say that with regard to this matter no Marxist physicists' voices were heard at the congress. Their "philosophical" speeches would have been most appropriate.

New quantum mechanics by Schrödinger is based on completely different premises. It is essentially a wave theory par excellence.

The questions on classical and molecular mechanics and on discontinuity and continuity were raised in debates on the Schrödinger theory. V. R. Bursian aptly compared the development of modern physics with the development of optics: the development of physical optics is characterised by the struggle between Newtonian emission (particle) theory and wave theory. And while several phenomena are well explained by one theory, it fails to explain a whole range of phenomena which are very well explained by the other theory. The question is not about an absolute authenticity of one theory or another but about their synthesis. We habitually think that our mechanics and its main laws derived from macro-world motion must be an absolute truth. But this conviction certainly is unfounded, and it is very likely that when we move to studying the micro-world we shall see that its laws are significantly different. The question is about the limit of the validity of the laws of classical physics. Both types of laws operate in nature, and it would be a mistake to imagine the laws of classical physics as a simple sum of molecular laws; laws of the micro-world fundamentally and qualitatively differ from laws of the macro-world.

In modern optics some phenomena are interpreted with the help of wave theory and some with the help of particles.

Schrödinger's mechanics seeks to provide a synthesis of both, and thus to build such a theory of matter and such a mechanics that would embrace both classical and molecular laws of motion. The Schrödinger theory also considers phenomena and laws of the micro-world to be different from those of the macro-world but recognises their unity. We now face the necessity of a synthesis of wave and particle theories in optics; similarly, the Schrödinger mechanics seeks to provide a synthesis of mechanics of wave and particle motion, a new theory of matter and a new mechanics that would embrace both classical and molecular laws of motion, i.e. would synthesise and generalise both Newtonian and quantum mechanics. In his mechanics Schrödinger arrives at differential equations which, as we know, require for their solution definite boundary conditions. Schrödinger assumes that the function giving the solution of this equation must be *continuous* in the entire space, therefore, he concludes, this equation is solvable not for all parameter values but only for certain *discontinuous* values. It turns out that these parameter values correspond to quantum numbers and in aggregate characterise an atomic state. Therefore, the Heisenberg table (matrix) is obtained from the Schrödinger equation and its formal symbols get physical meaning. And as we saw above both these theories arrive at the same quantum numbers that were considered by the old quantum theory.

A discrete aggregate of values that describe an atomic state is an aggregate of eigenvalues (Eigenwerte) for which the Schrödinger equation has a solution. And this discrete aggregate of values is obtained on condition that the function in the equation must be continuous at all points of space.

Thus, continuity and discontinuity do not exclude but imply each other. This was stated by professor Fock in his speech; he stressed that we must give up old concepts of discontinuity and continuity.

Y. I. Frenkel's talk "On results and prospects of quantum theory" also touched upon a very important and significant issue. The main peculiarity of quantum mechanics is that unlike ordinary mechanics, it deals with statistical collectives and not with single individuals (bodies). We do not observe the behaviour of a single atom but only that of a statistical aggregate of atoms. In this sense, we can draw an analogy between the methods of quantum mechanics and of the kinetic theory of matter. However, it seems to us wrong to draw here a further conclusion about the fundamental impossibility of observing individual states of an atom. The fact that the new quantum mechanics by its methods approaches statistical mechanics only suggests that laws of the statistical collective of atoms qualitatively differ from the laws of motion of an individual atom.

As we have seen above, the concept of a formal mathematical symbol is introduced instead of the idea of the position of an atom and its coordinates at a given moment of time; this symbol does not refer to a single atom at a given moment of time but characterises, as it were, its full "life history", the state in which it can be. The Heisenberg symbolic matrix may be interpreted either as corresponding to a statistical aggregate of atoms at a given moment of time or as a "life history" of a single atom. Heisenberg chose the latter. But we think it would be correct to interpret the matrix as corresponding to a statistical collective. Then it is reasonable to consider [also] the state of an individual atom.

Here we must finish our short notes on these exceptionally interesting developments in theoretical physics. To reiterate, we think that dialectical materialists cannot ignore the questions that are in the focus of modern physics. In turn, physicists would greatly benefit from the knowledge and use of the dialectical method. The discussions on new quantum mechanics took place at plenary meetings of the congress; however, no discussion in the true sense of the word took place, although here, undoubtedly, were a lot of fundamental questions in need of discussion, not to mention special issues (i.e. derivation of the Schrödinger equation). And the discussion might have taken place. Lack of comments in the debates cannot be considered an indicator of unanimity in the matters relating to the development of new physics.

While the debates demonstrated the breakdown of old views, the working group on general physics heard many talks attempting to derive new mechanical laws from Newtonian mechanics (K. N. Shaposhnikov) or give a physical interpretation to quanta based on the Thomson model (N. P. Kasterin). This approach demonstrated reasonably effective results but at the cost of adopting some new assumptions—postulates, [that are] so far arbitrary. Such attempts were marked with a certain tension. A pity that their authors did not participate in the "quantum" discussions at general meetings of the congress. Curiously, the working group on general physics heard a talk on the history of natural science: A. N. Schukarev saw in Boscovich[14] a predecessor of Bohr but this historical excursus is rather a stretch. Although the history of previous thought is of great and unprecedented importance for modern science, it requires a more objective and deeper approach. In this sense, for example, Newton's ideas on optics are much more instructive for the theory of light quanta.

It is really unfortunate that the theory of relativity has not been adequately reflected at the congress and some comrades tried to hasten its complete burial, albeit a little prematurely. Comrade A. K. Timiryazev's talk and his well-known point of view on the Dayton Miller experiments were met by a rather strong and it seems, valid objection from A. F. Joffe, who pointed out the crudeness and scientific incompetence of these experiments. According to A. F. Joffe who witnessed the experiments, Dayton Miller did not follow elementary rules of scientific experimentation. We would like to quote here just one of A. F. Joffe's thoughts, i.e. that although now the Dayton Miller experiments demonstrated results matching theory, the experimental conditions were so imperfect that it would not be surprising if these experiments yielded results several times exceeding the theoretical data. At the moment Millikan, a famous experimental physicist performs anew the Dayton Miller experiments, properly following the scientific technical rules. We have to wait for his results.

Relativity theory was mentioned only in this context. But the broadest methodological questions are to be raised in connection with this theory and also connecting it with both classical physics and quantum theory. In this respect comrade A. K. Timiryazev narrowed the issue and in his talk did not touch upon the general methodological and philosophical questions.

Among the talks that touched on the methodological and philosophical questions (apart from the most important ones above) one can mention L. Y. Strum's talk "Superluminal speed in the special theory of relativity and the causality principle", where the speaker linked the causality concept with the concept of irreversible processes, and B. L. Rosing's talk on advanced and retarded potentials, raising the question of the synthesis of "the present and the past". Unfortunately, due to the publication length constraints we will not dwell on these talks.

The development of theoretical natural science and the need for systematisation and generalisation of multiple experimental data will with necessity lead natural scientists to the dialectical approach to studying nature, i.e. dialectical materialism. More than once repeated by Engels, this thought came to mind especially often during the congress. The congress clearly demonstrated a strong "demand" for Marxist

---

[14]TN: Roger Joseph Boscovich (1711–1787), well-known in the history of science, was a Roman Catholic priest, originally from Dubrovnic, who studied in France and Italy. He was proficient in the science and mathematics of his day and highly creative. He developed an atomic theory, intended to explain physical phenomena, where the atoms were points with no extension and where the force between them could be attractive or repulsive, varying with distance. This approach gained some adherents in the eighteenth and nineteenth century.

methodology, although for the time being, probably an unconscious one on the part of physicists.

Russian physicists tend to have healthy ideas and not to have a deliberate and conscious scientific "reactionary character"; they follow the old materialistic tradition noted by Lenin in Russia. Marxists would be able to cooperate with most Russian scientists. This is the conclusion we made at the Fifth Congress of physicists.

Attendees at the Fifth Congress
(Maksimov is included but is not listed as attending)

Boris Hessen

I. Y. Tamm

A. F. Joffe

V. A. Fock

Y. I. Frenkel

V. R. Bursian

P. P. Lazarev

V. K. Lebedinsky

B. L. Rosing

A. N. Schukarev

K. N. Shaposhnikov

L. Y. Strum

P. S. Tartakovsky

A. K. Timiryazev

A. A. Maksimov

V. P. Egorshin

N. P. Kasterin

References

Bohr, N. (1926). Atomtheorie und Mechanik. *Die Naturwissenschaften [The Science of Nature]*, *14*(1), 1–10.

Bokulich, P., & Bokulich, A. (2005). Niels Bohr's generalization of classical mechanics. *Foundations of Physics*, *35*(3), 347–371.

Feynman, R. (1983). *QED, the strange theory of light and matter*. Princeton: Princeton University Press.

Heisenberg, W. (1926). Quantenmechanik. *Die Naturwissenschaften [The Science of Nature], 14*(45), 989–994.
Hertz, H. (1956). *Principles of mechanics*. London: Dover.
Joravsky, D. (2019). *Soviet marxism and natural science, 1917–1932* (2nd ed.). Abingdon, UK: Routledge.
Lenin, V. (1977). *Collected works* (Vol. 14). Moscow: Progress Publishers.
Mehra, J., & Rechenberg, H. (1982). *The historical development of quantum theory* (Vol. 1, Part 2). Heidelberg: Springer.
Pais, A. (1991). *Niels Bohr's times*. Oxford: Clarendon Press.

Chapter 3
On Comrade Timiryazev's Attitude to Modern Science

Boris Hessen and Vasilii Egorshin

We published our article on the Fifth Congress of Russian physicists in the first issue of this magazine and signed it H. E.[1]

Comrade Timiryazev was displeased with the article and particularly with our short report on debates related to his talk.

We do not consider the pages of our magazine to be an appropriate platform for polemics on specialised physics questions. "*Pod Znamenem Marksizma*" has different objectives. Therefore, we are not going to engage in a discussion on purely physical, empirical and technical aspects of the Dayton Miller[2] and Kennedy experiments.[3] Such a discussion should take place in specialised physics magazines. The section in Comrade Timiryazev's article dedicated to polemics and objections to Professor S. I. Vavilov[4] would be most relevant in the same magazine that published S. I. Vavilov's article.

We shall therefore focus on the fundamental issues raised in the article of Comrade Timiryazev.

It has become fashionable lately to blame Marxist Dialecticians for neglect of science and attempts to adjust science to dialectical "schemes" etc. The blame originated in the Mechanists' camp to which Comrade A. K. Timiryazev belongs. But even without additional explanations it should be clear that this blame by the Mech-

---

[1] TN: See Chap. 2.

[2] TN: For Dayton Miller see Appendix.

[3] TN: Roy J. Kennedy was an American physicist who also carried out experiments relating to the speed of light.

[4] TN: On Vavilov see Appendix.

TN: Translated from *Pod Znamenem Marksizma* (*Under the Banner of Marxism*), 1927, Nos. 2–3. Cited by Joravsky (2019), p. 186, n. 360 and Josephson (1991), p. 385, n. 75.

TN: For more on Egorshin and Timiryazev see Appendix.

B. Hessen (✉) · V. Egorshin
Abingdon, UK
e-mail: chris.talbot@protonmail.com

C. Talbot and O. Pattison (eds.), *Boris Hessen: Physics and Philosophy in the Soviet Union, 1927–1931*, History of Physics,
https://doi.org/10.1007/978-3-030-70045-4_3

anists contains an obvious contradiction. Indeed, dialectical materialism is a much wider worldview than mechanical materialism. Not only, as noted by Lenin, multiple scientific discoveries of the last decades demonstrated the whole insufficiency of the narrow mechanical worldview, the same discoveries confirmed dialectical materialism in the best possible way. Lenin showed also that the lack of dialectical knowledge leads many natural scientists to idealism, Machism and other "schools and mini-schools". Therefore, it would be quite wrong when criticising and refuting many idealistic perversions of science to go back to the prevailing views of 50 years ago. New dialectical natural science must synthesise the materialistic worldview and the new *facts* of the latest science.

Mechanists hold a different view. By denying dialectics (only saying that they accept it is not enough!) they must simply ignore and indiscriminately deny everything that does not fit in the Procrustean framework of the Mechanist's understanding of the world. Anyone can see that this framework is incomparably narrower and more lopsided than the dialectical teaching. This is brilliantly expressed by Lenin in his extract on dialectics.

> "Dialectics as *living*, many-sided knowledge (with the number of sides eternally increasing), with an infinite number of shades of every approach and approximation to reality . . . here we have an immeasurably rich content as compared with "metaphysical" materialism . . ." Even "Philosophical idealism is *only* nonsense from the standpoint of crude, simple, metaphysical materialism. From the standpoint of *dialectical* materialism, on the other hand, philosophical idealism is a *one-sided*, exaggerated, überschwengliches [effusive] . . . development (inflation, distension) of one of the features, aspects, facets of knowledge, into an absolute." And further "Idealism is clerical obscurantism. True. But philosophical idealism is ("*more correctly*" and "*in addition*") a *road* to clerical obscurantism through *one of the shades* of the infinitely complex *knowledge* (dialectical) of man". (Lenin's italics).[5]

Therefore, blame in trying to ignore science and to fit it into narrow and rigid schemes, if blame we must, lies only with Mechanists and not the Dialecticians.

And this article by Comrade A. K. Timiryazev is the best confirmation of the above. He believes that relativity theory and quantum theory cannot get along with materialism and finds both these theories 100% Machist.

Comrade Timiryazev writes, "For most modern theoreticians and especially for the Russian ones 'a purely descriptive philosophy' is still the only scientific philosophy."

If we look at the whole body of literary works by A. K. Timiryazev we shall find that he does not see any other enemy of materialism apart from "a purely descriptive philosophy". For example, Kantianism seemingly does not exist for him. However, after Helmholtz, who is much respected and highly regarded by A. K. Timiryazev as an exemplary dialectical materialist, the Kantian trends among theoretical natural scientists must not be discounted.

---

[5] BH and VE: Lenin, V. I., "On the question of dialectics", *Pod Znamenem Marksizma* (*Under the Banner of Marxism*) 5–6 (1925): p. 17. TN: English translation taken from *Philosophical Notebooks*, (Lenin 1976), pp. 360–361.

But this is in passing and so let us assume that there is nothing to add here and let us focus only on the Machist "purely descriptive philosophy". What does this have to do with relativity theory? A. K. Timiryazev answers this question very clearly and unequivocally.

He states that "relativity theory and quantum theory have brought about a new outbreak of Machism". This is hard to contest. But this does not justify a devil-may-care attitude to these two theories and an indiscriminate denial of both. In his approach A. K. Timiryazev does not follow Lenin, who in his *Materialism and Empirio-criticism* (Section: "The Essence and Significance of 'Physical' Idealism")[6] gave a remarkable analysis of Mach's, Duhem's, Stallo's and, in passing, Rey's dodging and weaving.[7]

Although the most ruthless enemy of Machism, Lenin did not indiscriminately deny certain progressive ideas of the physicists Duhem and Stallo. According to Lenin, "they are in reality vacillating between idealism and *dialectical materialism* . . ." (p. 316, 1920 edition,[8] italics here and below by H. and E.). Lenin says, "[they] most energetically combat the atomistic-*mechanical* conception of nature. They show the narrowness of this conception, the impossibility of accepting it as the limit of our knowledge, the *rigidity* of many of the ideas of writers who hold this conception. And it is indeed undeniable that the *old* [Lenin's and BH and VE's italics] materialism did suffer from such a defect . . ." Seems undeniable.

Comrade Timiryazev likes to quote Lenin but you will find no quotations similar to the one above in his works. Comrade Timiryazev hides behind Lenin's broad back and shoots from the hip at the most eminent modern scientists and their theories and labels them with similar monotonous grey names "scientific fashion", "fashionable theory", "fanatical fans" etc.

According to Comrade Timiryazev, the Marxist approach in modern natural science must be reduced simply *to turning back* from Planck, Sommerfeld and Bohr, etc., i.e. the "Machists-Einsteinists", not to mention Einstein himself, to Helmholtz, Maxwell, Thomson and Boltzmann. Comrade Timiryazev believes that by closely following these scientists one would not retreat a single step from consistent (and therefore dialectical) materialism. It is from them, he suggests, all Marxists should learn both materialism and dialectics.

Comrade Timiryazev forgets a "small" detail: although the old teachers might have been materialists, they might not have been dialectical materialists but metaphysical ones (certainly with some elements of spontaneous dialectics that are inherent

---

[6]TN: Chap. 5, Sect. 8 in Lenin (1977).

[7]TN: For Ernst Mach, see Appendix. Pierre Duhem (1861–1916) was a French theoretical physicist but today is well-known for his contribution to the history and philosophy of science. John Stallo (1823–1900), a German–American intellectual who was an early contributor to the philosophy of science to whom Mach was sympathetic. Lenin attacks Duhem and Stallo, like Mach, for alleged "idealism". Abel Rey (1873–1940) was a French philosopher and historian of science whose philosophy Lenin denounced as "positivism" but whose summary of the latest developments in physics Lenin studied as being made "carefully and in general conscientiously" (p. 254).

[8]TN: References to the Russian edition. The English translation is in Lenin (1977), p. 310.

in anyone, even the "Black-Hundred" Einstein).[9] Meanwhile Lenin, whose words Comrade Timiryazev does not really favour with his attention, wrote in the same chapter as above, "The basic materialist spirit of physics, as of all modern natural science, will overcome all crises, (Lenin speaks here of Machism—H. and E.) but only by the indispensable replacement of metaphysical materialism by dialectical materialism." (p. 132 in the above quoted publication,[10] italics by H. and E.). Let the reader judge whether this agrees with the *downright nihilism* and the stubborn "old belief"[11] of Comrade Timiryazev.

The Heisenberg theory (of quantum mechanics) was very popular last year, you understand, because it was "new" and "in vogue"; and now an even newer Schrödinger's theory is equally popular for the same reason.[12]

There is an impression that all physicists are like flirty Paris fashionista who change their attitudes and physical theories as often as the representatives of the fair sex change their clothes. And the more Machist is a theory, the more greedily physicists jump at it.

Reality is not quite like this or rather, quite unlike this. Indeed the Heisenberg theory spread widely among physicists, but not because they are so susceptible to a new Machist front-pager (although individual lovers of this type of brouhaha surely could be found), but because the Heisenberg theory in spite of its formal character is a step forward in atomic mechanics, as a whole range of experimental data fits into the framework of this theory while it is inexplicable from the viewpoint of classical quantum theory.

But Heisenberg's theory is a formal description! It wants to deal only with the "observable-in-principle values". It refuses to build any models! What good can a theory like this do apart from pouring water on the mill of the "purely descriptive philosophy"? This is the reasoning of Comrade Timiryazev.

A couple of words about description. Is every description always identical to Machism? We believe this question does not have a very simple answer. An "economical description" as a general methodological and *epistemological* principle is one thing, but a description as a *phase* in physics research is another matter.

---

[9]TN: The Black Hundreds were an ultra-nationalist movement in Russia in the early twentieth century. It is used here as a term of abuse, meaning "reactionary", as Einstein was originally a supporter of Mach.

[10]TN: i.e. the Russian edition of "Materialism and Empirio-criticism". The English translation given here is Lenin (1977), p 306.

[11]TN: In the context of Russian Orthodox Church history, the Old Believers, or Old Ritualists separated after 1666 from the official Russian Orthodox Church in protest against church reforms introduced by Patriarch Nikon of Moscow between 1652 and 1666.

[12]BH and VE: Here is a correct reference: Comrade Timiryazev writes that the Schrödinger theory appeared in the autumn of 1926. As a matter of fact, Schrödinger's first works appeared in February 1926.

As is known, Mach believed Kirchhoff to be like-minded and supportive of "economical description". In his lectures on mechanics Kirchhoff defines the objective of mechanics as follows, "Mechanics sets tasks which I define as *description* (original italics) of motions taking place in nature; indeed, the complete and simplest description".[13]

Nevertheless, Lenin defends Kirchhoff against Mach! When in *Knowledge and Error* Mach says that his economical description principle and "Kirchhoff's complete and simplest description . . . with slight variations, express one and the same thought", Lenin passionately intercedes for Kirchhoff, "Is this not a model of confusion? 'Economy of thought', from which Mach in 1872 inferred that sensations *alone* exist . . . is declared to be *equivalent* (italics by Lenin) . . . to the simplest *description* (of an *objective reality*, the existence of which it never occurred to Kirchhoff to doubt!)" (v. 10, p. 139).[14]

Therefore, there are descriptions and descriptions!

The whole point is that "economical description" according to Mach is an epistemological principle which he considers as a corner stone in the theory of cognition. Kirchhoff regards "complete and simplest description a way to remove from mechanics those concepts whose physical meaning is obscure, and first of all the concept of force."[15]

In his matrix method Heisenberg strives to overcome a whole range of fundamental and principal difficulties in the development of the mechanics of atoms. These difficulties arise from the imperfection of our models and ideas about atomic processes. The Heisenberg theory successfully, albeit formally, overcomes some of these difficulties.

Therefore, in our report we on the one hand, pointed out the possibility of deriving Machist conclusions from principles of the Heisenberg theory (Heisenberg himself arrives at these conclusions), but on the other hand, we thought it imprudent to reject

---

[13]BH and VE: (Kirchhoff 1897). TN: The translation from the Russian text is given here. It is apparently intended to correspond to the original German sentence in Kirchhoff (1897), p. 1: "Die Mechanik ist die Wissenschaft von der Bewegung; als ihre Aufgabe bezeichnen wir: die in der Natur vor sich gehenden Bewegungen "*vollständig*" und "*auf die einfächste Weise*" zu beschreiben." This translates as "Mechanics is the science of movement; we define as its task: to describe completely and in the simplest manner the movements taking place in nature." The authors italicising of "description" does not correspond to the original.

[14]TN: The English translation is given here, (Lenin 1977), p. 171.

[15]TN: The authors are presumably referring to Kirchhoff (1897). The translation from the Russian text is given here. Although given in quotes it appears to be attempting to summarise Kirchhoff's view on force (Kraft) rather than giving an exact quotation. To suggest that Kirchoff wants to "remove" the concept of force from mechanics is not really correct, rather he wants to remove obscure definitions referring to force. In the Preface he writes that "One tends to define mechanics as the science of the forces, and the forces as the causes which produce, or strive to produce, motions . . . it seems to me [to be] desirable to remove from it such obscurities, even if that is only possible by narrowing its task." German original: "Man pflegt die Mechanik als die Wissenschaft von den Kräften zu definiren, und die Kräfte als die *Ursachen*, welche Bewegungen hervorbringen oder hervorzubringen *streben* . . . scheint es mir wünschenswerth, solche Dunkelheiten aus ihr zu entfernen, auch wenn das nur möglich ist durch eine Einschränkung ihrer Aufgabe.".

the *physical* content of this theory only on this basis. We believe this to be the only correct approach to the Marxist appraisal of different physical theories. Description will never be a bogey for us. We need to understand what kind of a description it is. The Balmer formula was also just a description at the time but did it not signify a step forward? Bohr came after Balmer and explained [to us] the physical meaning of the formula and of its constants (the Rydberg constant).[16] Why should we reject Heisenberg's theory outright? It might also receive its physical explanation.

Comrade Timiryazev says, "Machism flourishes in the areas where we lack knowledge and where we are forced to be temporarily limited by a formal description. And Machism considers this situation to be the final solution." Undoubtedly a formal description is more popular in the areas where we fail to explain classical laws on the basis of molecular ones. But what shall we do if such an explanation is as yet impossible at the current stage of scientific development?

Shall we give up a (mathematical) formulation of general laws whose molecular mechanism is unknown because it is a formal, Machist and pure description?

Let us hear a relevant opinion of one physicist whom Comrade Timiryazev can hardly accuse of Machism:

In certain cases, he says, we know all changes in detail.

> Thus the motion of the moon may be described by stating the changes in her position relative to the earth in the order in which they follow one another.
>
> In other cases we may know that some change of arrangement has taken place, but we may not be able to ascertain what that change is.
>
> Thus when water freezes we know that the molecules or smallest parts of the substance must be arranged differently in ice and in water. We also know that this arrangement in ice must have a certain kind of symmetry, because the ice is in the form of symmetrical crystals, but we have as yet no precise knowledge of the actual arrangement of the molecules in ice. But whenever we can completely *describe* the change of arrangement we have a knowledge, perfect so far as it extends, of what has taken place, though we may still have to learn the necessary conditions under which a similar event will always take place.

These were the tasks of physics, according to James Clark Maxwell,[17] in the areas where the molecular mechanism was unknown.

In modern physics, we, on the one hand, are more or less familiar with the molecular mechanism that allows us to explain the laws of classical physics in the area of the kinetic theory of matter, but on the other hand, we have no similar explanation in the area of electromagnetic phenomena.

Most physicists would probably agree that it would be great to be able to construct a full picture of all electromagnetic phenomena on the basis of an ether with strictly

---

[16]TN: The Balmer series refers to lines at certain frequencies in the spectrum of light of a hydrogen atom. Rydberg gave a formula to calculate the frequencies of such lines. The formula for hydrogen contains a constant named after him.

[17]BH and VE: Maxwell, "Matter and motion", Russian translation, p. 1. TN: Italics added by BH and VE. The English original given here is Maxwell (1876), or https://archive.org/details/mattermotion00maxwiala/page/n8/mode/2up, pp. 1–2.

defined qualities. Unfortunately, by following this approach physics encountered extraordinary difficulties; and these difficulties resulted in the choice of a second approach pointed out by Maxwell.

Here is just one historical example that demonstrates the scope of the difficulties in the construction of a theory of electromagnetic phenomena based on a certain model of the ether.

William Thomson, one of the greatest physicists of the nineteenth century, dedicated his whole life to constructing a theory of electromagnetic phenomena from the ether. This is what this great physicist, a contemporary of Faraday, Helmholtz and Maxwell, said at the celebration of the fiftieth anniversary of his scientific work, "There is but one word that would describe my hard efforts to advance science over the 50 years—failure. I know now about the electric and magnetic forces and the interaction between ether, electricity and ponderable matter no more than I knew and tried to convey to my students 50 years ago, when I started as a professor."[18]

When we speak of a necessity of "a description" we by no means want to say "ignorabimus"[19] with regard to ether but want to stress that the situation is not as simple as Comrade Timiryazev depicts. These difficulties should be considered and analysed. If instead, we should consider Machism and unwillingness to accept the works by J. J. Thomson[20] to be the only reason for the formal approach, we shall not advance the solution of physical problems.

These are the criteria we use in appreciation of the theories of Heisenberg and Schrödinger.

Let us now turn to the question of relativity theory. Here Comrade Timiryazev holds such an extreme and one-sided view that he is forced to issue many promissory notes to justify it, and we appreciate his persistence and firmness.

Comrade Timiryazev maintains the point of view that relativity theory is absolutely incompatible with materialism. It is permissible, then, to ask what if in spite of everything, experiments would confirm the *physical* aspect of relativity theory (while Lenin already said everything there was to say regarding the nonsensical aspects of philosophical and idealistic relativism)? What if the *physicist* Einstein turns out to be right, what then? Will the materialistic worldview collapse altogether? Will idealism and clericalism take its place? *Such* an understanding of materialism—as connected with Newton's mechanics—would present a very low opinion of materialism indeed. In this case the worldview will change (from materialism to idealism) depending on the new and, in their turn, transitory scientific theories.

---

[18]BH and VE: Silvanus P. Thompson, "Life of Lord Kelvin, vol. 2," p. 1073. TN: (Thompson 2005), pp. 1072–3.

[19]TN: Latin, first person plural future active indicative of ignoro ("we won't know").

[20]TN: Sir Joseph John "J. J." Thomson, (1856–1940), famous British physicist. He discovered the electron, for which he received the Nobel Prize. He also discovered isotopes, and invented the mass spectrometer.

But this is precisely Comrade Timiryazev's view on materialism. In the article published here we read, "Comrade H. E.[21] is perplexed as to why the speaker has not included questions regarding the philosophy of relativity theory? Mostly because the *whole point* now is whether the conclusions by Dayton Miller are proved correct. *Depending on the answer* we shall either return to healthy *materialism* and to discovering new forms of matter or new forms of moving matter in this area of physics, or we shall continue to flounder in the Machist sea of 'pure mathematical description'". (Italics by H. and E.)

Either materialism—and then no relativity theory, *or* relativity theory will be proved, and then Machism celebrates victory. This dilemma is put to [a wide mass of] readers.

In contrast to this we follow Lenin and Engels in thinking that "materialism must change its form with every epochal discovery made in natural science or history" ("not to mention the history of mankind").[22] Although Einstein's theory at present is not as yet strictly proved by experiments, the Marxist methodology has to wait for the final solution of the physical problem obtained by physical methods; and every attempt to use this problem (both before and after the experimental confirmation) made by the Machist, Fictionalist[23] or any other idealistic philosophy must be decisively stopped. If it is confirmed that matter moves following Einstein's laws and not Newton's laws, matter would still be matter as an objective reality, and Comrade Timiryazev's statement that inevitably in this case "matter disappears, and only equations are left" is pure fantasy.

Idealism once used the electron theory for its own purposes; this theory allegedly contradicted materialism. Now then, what would you say about a "materialist" who during the experiments with electromagnetic mass would have been shouting at every corner, "Help! Materialism is dying, it is hanging by a thread!" This would be truly obscurantism. Lenin the materialist chose a diametrically opposite view. He was not afraid to "review" Engels's *form* of materialism in order to clarify the definition of matter; thus, the electron theory turned out to be not a refutation but a brilliant confirmation of dialectical materialism.

If Lenin the philosopher had to delve into the essence of new scientific discoveries in order to assess them, then Marxist natural scientists must consider it their primary duty. They must study the *physical aspect* of every new theory and only upon testing it in this manner, accept or reject the theory. Concerning Comrade Timiryazev's approach to relativity theory, it could be called *nihilistic*, and it has been such for a long time irrespective of the Dayton Miller experiments.

We believe it is too early for Marxists to pronounce their judgment on relativity theory; in our article on the congress of physicists we have adopted neither A. F. Joffe's view, nor that of A. K. Timiryazev. We think that physics does not yet

---

[21]TN: "H. E." is the author's name given in the article in Chap. 2, obviously referring to Hessen and Egorshin.

[22]TN: *Ludwig Feuerbach* (Engels 1990), pp. 369–370, (Lenin 1977), p. 251.

[23]TN: Fictionalist: Reference to the German Kantian philosopher Hans Vaihinger's "Philosophy of As If." See (Vaihinger 2000).

possess final experimental results, i.e. experimentum crucis. However, we believe that Newton's mechanics is undoubtedly insufficient regardless of the nature of its further add-ons, and here we differ with Comrade Timiryazev. Whether it will be supplemented by Einstein's theory or some other theory will be up to physics, but none of them will invalidate materialism.

We think that this view is shared by most physicists-materialists who are *now* supportive of relativity theory, and whom Comrade Timiryazev with an amazing ease, labels idealists simply on account of relativity theory. For example, take Planck. This famous physicist has strongly fought Machism for many years, but now, according to Comrade Timiryazev, turned out to be Machist. However, M. Planck recently wrote,

> Certainly the last decision on the validity and meaning of relativity theory belongs to experiment, and the very possibility of its experimental verification is the most important testimony to the fruitfulness of a theory. So far, no contradiction with experiment has been established. I would like to emphasize this fact in particular in opposition to certain news that has recently reached the general public. But even those who for some reason believe and think the occurrence of a contradiction with experiment to be probable, in their own interest can do no better than cooperate in the development of relativity theory and its further results. For this will be the only way to experimentally refute relativity theory.[24]

Any Marxist physicist can but agree with this approach. Where does Comrade Timiryazev find Machism here? The same reasoning is to be found in the article by academician A. F. Joffe in *Pravda* of 1 January this year, which Comrade Timiryazev has already labelled idealistic and Machist.[25]

Comrade Timiryazev may certainly reject the physical aspect of relativity theory, too. It is not improbable that relativity theory in its present form will be refuted. However, in this case Comrade Timiryazev should defend his ideas first and foremost among physicists. But we have not seen his single article in the specialised physical journals that put forward his view. Comrade Timiryazev obviously believes that his essays in the magazine *Pod Znamenem Marksizma* and in *Pravda* will bring him victory over all "Machist" physicists. We think this belief is unfounded.

---

[24]BH and VE: M. Planck, "Physical laws in the light of new research", Russian translation in *Uspekhi Fizicheskikh Nauk*, 6 (1926), p. 192. TN: German original, (Planck 1926). The translation closely follows the original.

[25]BH and VE: A. F. Joffe says, "A theory describing material phenomena and physical processes in matter cannot contradict the materialistic worldview if only it aspires to describe qualities of matter in the best possible way." What is the specific content of a physical theory is a "question of expedience" (*Pravda*, 1 January 1927).

Regarding this paragraph A. K. Timiryazev comments as follows, "He (A. F. Joffe) definitely equates materialism and Machist philosophy ...with just one word economical substituted by the word expedient" (*Pravda*, 26 February 1927). A. K. Timiryazev is surprisingly generous with the label "Machism". A. F. Joffe everywhere speaks of matter, but Comrade Timiryazev sees here a "complex of sensations".

A. F. Joffe uses an unfortunate expression "expediency". However, the whole context demonstrates whether it expresses a Machist "description" or not. A. F. Joffe's next sentence is, "The best of these ideas (of matter) is the one that closer than others approximates the qualities of real matter." Consequently, A. F. Joffe uses "expediency" in the sense of conformity to reality.

Therefore, here is an obviously inappropriate accusation of Machism that counts on readers being too lazy to look up A. F. Joffe's article.

Indeed, Comrade Timiryazev's paper at the congress of physicists was such an attempt. Firstly, however, this attempt was the only one, and secondly, even at the congress it remained insufficiently implemented, if only because the paper was read in a working group (five working groups at the congress met simultaneously) and not at a plenary session.

If Comrade Timiryazev believes he should seriously fight against relativity theory, he should not ignore the issue of defending his views among physicists.

Let us now turn our attention to another problem, which Comrade Timiryazev touched upon in his article aimed against us, i.e. the new quantum mechanics.

"The Schrödinger theory in its methods is as formal as the Heisenberg theory. According to this theory waves that do not possess a material carrier, form electrons, i.e. matter. It is a classical illustration of Lenin's words, 'an attempt to conceive motion without matter'."[26]

To begin with, firstly, nowhere in our last article have we affirmed that Schrödinger's theory in its present form is unreservedly acceptable to dialectical materialism. Both classical physics and the old quantum theory turned out to be unable to explain nature. We see new theories attempting to avoid the dead end. Some of them might be unsuccessful but in no way does this fact eliminate the *collapse* of the old mechanical physics.

Furthermore, secondly, we have not denied the formal character of Schrödinger's theory. We also have pointed out the necessity for a further discussion regarding the derivation of Schrödinger's equation.[27] Most importantly, we have stressed that Schrödinger's theory tries to present a synthesis of classical and molecular mechanics. We have pointed out the difference between Schrödinger and Heisenberg and have noted the unacceptability of some *fundamental* premises of Heisenberg that led him to doubt the degree of reality of electrons. Schrödinger's methodological premises are of a different character.

As we have pointed out, his [Schrödinger's] task is "to build such a theory of matter and such a mechanics that would embrace both classical and molecular laws of motion, i.e. a synthesis of both Newtonian and quantum mechanics" (p. 139, *Pod Znamenem Marksizma*, no. 1).[28]

Comrade Timiryazev actually admitted that Newton's mechanics is insufficient and such a synthesis is necessary, when he wrote in his article "Quantum theory and modern physics" the following: "The task is reduced to linking the new to the old. We have to find the limits of applicability of classical mechanics and electrodynamics and to figure out the common laws that should unite them and add to them, as it is clear that the 'quanta' adds something new" (Collection: *Natural science and dialectical materialism*, p. 138).[29]

[26]TN: Lenin criticises "to conceive motion without matter" as idealist. Lenin (1977), pp. 266–273.

[27]TN: The authors refer to the article in Chap. 2, pp. 20–21.

[28]TN: i.e. Chap. 2, p. 20.

[29]TN: The authors are referring to A. K. Timiryazev's article in the collection: *Natural science and dialectical materialism* (Moscow: Materialist Publishers, 1925).

Thirdly, we have already read about some attempts to give a clear interpretation of Schrödinger's equation using a material carrier; they resulted in a widening of our understanding of electrons, and this understanding is very close to the one defended by Comrade Timiryazev.

The problem of Schrödinger's theory is not so simple to solve!

There have been many strange and incomprehensible concepts in classical quantum theory, and primarily the quantum conditions that were expressed in whole numbers.

Comrade Timiryazev in his article "Newest attempts to resurrect teleology in physics"[30] accused Bohr of only partly overcoming Mach because his theory does not explain the electron jump mechanism and the nature of quantum conditions. Schrödinger's theory attempts *to explain* the discrete aggregate of whole numbers in quantum conditions by regarding wave processes as a basis of mechanics. This is the only way to understand why whole numbers come up in quantum theory.

This is also how Comrade Timiryazev explains this fact as becomes evident from his article "Quantum theory and modern physics". Therefore, in this respect Schrödinger's theory is a step forward in *explaining* phenomena. For example, W. Wien[31] also supports this explanation. Here is what he said in his speech in 1926,

"Quantum theory's complete break from the old physical theories was particularly strange, as were its incomprehensible whole numbers and inability to make sense of the processes that actually took place. This is why Schrödinger made an effort to consider the whole problem as a vibration problem and thus make it more obvious and easier to understand."[32]

Yes, but this is conceiving motion without matter![33]

This is perhaps a rather hasty conclusion. We have earlier pointed out the difficulties encountered by physics with regard to the concept of the ether. Schrödinger derived his equation based on formal analogies between geometrical and physical optics. But does this mean that this theory fundamentally and methodologically does not allow a material carrier of oscillatory processes? The following facts demonstrate that it is not so: Madelung,[34] for instance, tries to give a 'clear interpretation of Schrödinger's equation.'[35]

Essentially, he believes that an electron is regarded as a liquid that continuously fills the entire space with a certain density and current at a certain speed. Madelung says, "Consequently, the quantum theory of stationary states of an atom is reduced to the hydrodynamics of continuously distributed electricity. In the case of several electrons in an atom we should accept that they mutually penetrate each other but

---

[30]TN: Ibid. p. 321.

[31]TN: William Wien (1864–1928), a leading German physicist. He is best known for his work on heat (black-body) radiation for which he won the Nobel Prize in 1911.

[32]TN: (Wien 1926), p. 14. The translation here closely follows the German original.

[33]TN: See p. 38, n. 26 above.

[34]TN: Erwin Madelung (1881–1972), a German physicist who worked on atomic physics and quantum mechanics gave an alternative "hydrodynamical" interpretation of Schrödinger's equation.

[35]BH and VE: (Madelung 1926a, b).

do not merge."[36] By no means should this explanation be considered final, but it demonstrates that Schrödinger's theory does not necessarily conceive motion without matter. In addition, this interpretation introduces a new view of electrons that undoubtedly widens our understanding. And this understanding reminds us of that of . . . J. J. Thomson, whom Comrade Timiryazev calls an exemplary materialist.

Here is what the latter wrote,

"Indeed, when we calculate the electromagnetic mass of an electron, or according to Thomson, 'mass of the ether connected with the lines of force of an electron', we must take into account the whole 'related' mass in the entire infinite space. Although most of this mass is in the immediate proximity of the electron, we have to sum up or 'integrate' over the entire space regardless of whether we believe in the existence of the ether or of 'the vacuum with electromagnetic properties', favoured by the supporters of 'pure description' and by the enemies of 'materialistic metaphysics'. Therefore, strictly speaking the carrier of mass of an individual electron is the entire world! Clearly under these conditions the old concept of impenetrability is sufficiently relative. Further, according to J. J. Thomson, we perceive the ether mass as weight only while it relates to the electric charge lines of force.[37] The rest of the mass for us has no weight: neither have we any impact on it, nor vice versa does it have any effect on us."(A. Timiryazev, Collection: *natural science and dialectical materialism*, p. 215).[38]

How far is this interpretation from Madelung?

For this reason alone, Comrade Timiryazev should not dismiss Schrödinger's theory as a "new vogue".

Very recently we received other interpretations of Schrödinger's theory, e.g. by Darwin in *Nature* dated 22.02.1927 (C.G. Darwin, The electron as a vector wave).[39]

As to Schrödinger, his views on the material carrier of wave motion in principle coincides with the views of Madelung.[40]

By no means do we affirm that Madelung's and Schrödinger's interpretation is the final concept of material substrate. Possibly here much will undergo a radical change. But the fact remains, Schrödinger's theory does not exclude a material carrier of motion. We note that Madelung published his interpretation in November 1926, therefore, it is quite wrong to say in March 1927 that Schrödinger's theory is a 'classic example of conceiving motion without matter'.[41]

And finally.

---

[36]TN: (Madelung 1926a), p. 1004.

[37]TN: Thomson developed a theory of "ether mass" in which the electromognetic field contributed to the mass of a charged particle, an approach which was eventually incorporated in Einstein's theory of special relativity. Thomson's initial paper in 1881 was Thomson (1881).

[38]TN: See p. 39, n. 30.

[39]BH and VE: (Darwin 1927).

[40]BH and VE: Compare (Schrödinger 1926). Particularly, paragraphs 2 and 7.

[41]BH and VE: It is interesting to note that Comrade Timiryazev, who considers Heisenberg's and Schrödinger's theories unacceptable for materialistic physics, has never spoken on this matter in the discussions that took place at the congress of physicists.

Comrade Timiryazev writes, "In their article H. E. praise Russian theoreticians for following Schrödinger and refusing to accept the formal theory of Heisenberg." As any unbiased reader would see in our article on the congress, we "praised" Russian theoreticians not for following Schrödinger, but for the fact that most speakers disassociated themselves from Machist tendencies that are present in Heisenberg's works. Not the same thing, it would seem. Comrade Timiryazev begins to read the unwritten.

However, if he wants to pose a fundamental question on Marxists' attitude to Russian physicists, here are two possibilities: either Comrade Timiryazev agrees with us in his assessment of Russian physicists who "do not have a deliberate and consciously scientific 'reactionary character'"; and in that "Marxists would be able to cooperate with most Russian scientists". In this case his ironic comments regarding "praise", etc. are totally incomprehensible and excessive. Or Comrade Timiryazev dislikes our conclusion regarding cooperation between Marxists and Russian physicists who spontaneously follow a materialistic tradition. In this case Comrade Timiryazev should clearly and unambiguously declare that he considers the main nucleus of Russian physicists reactionary, absolutely alien to us and unable to accept not only the ideas of modern materialism, but also the tasks of Soviet socialist construction.

Here we arrive at the conclusion where theory is intertwined with practical politics; here together with the Party and with the Soviet power we stick to the firm opinion that we *must* cooperate with the representatives of modern science, and we can make them dialectical materialists only through *cooperation* with them. The nihilistic view point of a certain scientific "Napostovstvo"[42] would cause irreparable damage to the revolution and to Marxism.

References

Darwin, C. G. (1927). The electron as vector wave. *Nature, 119*, 282–284.

Engels, F. (1990). *Marx Engels collected works* (Vol. 26). Moscow: Progress Publishers.

Joravsky, D. (2019). *Soviet Marxism and natural science, 1917–1932* (2nd ed.). Routledge, Abingdon, UK.

Josephson, P. R. (1991). *Physics and politics in revolutionary Russia*. Berkeley: University of California Press.

Kirchhoff, G. (1897). *Vorlesungen über Mechanik* (4th ed.). B. G. Teübner, Leipzig.

Lenin, V. (1976). *Collected works* (Vol. 38). Moscow: Progress Publishers.

Lenin, V. (1977). *Collected works* (Vol. 14). Moscow: Progress Publishers.

Madelung, E. (1926a). Eine anschauliche Deutung der Gleichung von Schrödinger. *Die Naturwissenschaften (The Science of Nature), 14*(45), 1004.

Madelung, E. (1926b). Quantentheorie in Hydrodynamischer Form. *Zeitschrift für Physik, 40*(3–4), 322–326.

Maxwell, J. C. (1876). *Matter and motion*. London: Society for Promoting Christian Knowledge.

[42]TN: Napostovstvo (Onguardism)—a group of proletarian writers who published a magazine Na postu in 1923–1925. According to Joravsky (2019), pp. 159–160, in literary disputes they were a tendency that wanted "to reject 'bourgeois culture' indiscriminately and to advocate unreasoning force in promoting 'proletarian' culture.".

Planck, M. (1926). Physikalische Gesetzlichkeit im Lichte neuerer Forschung. *Die Naturwissenschaften (The Science of Nature), 14*(13), 257.
Schrödinger, E. (1926). Quantesierung und Eigewertproblem, (Vierte Mitteilung). *Annalen der Physik, 81*(18), 109–139.
Thompson, S. (2005). *Life of Lord Kelvin* (Vol. 2, p. 1910). Reprint of Macmillan: American Mathematical Society—Chelsea, Providence, Rhode Island.
Thomson, J. J. (1881). On the electric and magnetic effects produced by the motion of electrified bodies. *Philosophical Magazine, 11*(68), 229–249.
Vaihinger, H. (2000). *The philosophy of "As if": A system of the theoretical, practical and religious fictions of mankind.* Routledge, London. Originally published in German in 1911.
Wien, W. (1926). *Vergangenheit.* Gengenwart und Zukunft der Physik: Maxheuber Verlag, Munich.

Chapter 4
On the Bicentenary of Isaac Newton's Death. Foreword to the Articles by A. Einstein and J. J. Thomson

Boris Hessen

A large number of global celebrations, speeches, talks and articles marked the bicentenary of Newton's death. Newton's work and thoughts have acquired special significance against the background of the revolution currently taking place in modern exact natural science. Many of Newton's ideas and particularly the ideas in his "Optics" have regained significance and renewed interest. At the same time, the weak aspects of his "Natural Philosophy" are particularly highlighted by the current crisis in modern physics. The main methodological concepts and premises of the system of Newton's physics must be revised. Indeed, theoretical natural science arrives at this revision by the very course of its development. How should the revision and further development of Newton's physics be carried out? How do the new concepts suggested by the development of physics over the last twenty years correspond with Newton's physics, if at all?

All these questions certainly arise irrespective of the bicentenary, but the latter has substantially stimulated interest in such questions.

The articles below by Einstein, J. J. Thomson and H. Lamb[1] follow different approaches to the [question of the] role and significance of Newton and the appreciation of his works.

Newton's celebrations in England were marked by particular splendour. The most prominent members of the Royal Society and physicists of England dedicated their

[1]TN: Translations into Russian of the following articles follow in *Pod znamenem marksizma*, pp. 166–186: Einstein (1927)—an English translation appeared in the Smithsonian Annual Report for 1927 (see https://www.pbs.org/wgbh/nova/article/einstein-on-newton), Thomson (1927) and Lamb (1927).

B. Hessen (✉)
Abingdon, UK
e-mail: chris.talbot@protonmail.com

C. Talbot and O. Pattison (eds.), *Boris Hessen: Physics and Philosophy in the Soviet Union, 1927–1931*, History of Physics,
https://doi.org/10.1007/978-3-030-70045-4_4

speeches to Newton's work. Thomson, Jeans, Lamb, Glazebrook[2] and many others outlined Newton's achievements in physics, mechanics, astronomy and mathematics.

However, these speeches although sometimes brilliantly worded almost completely lacked any kind of generalisation. Almost no attempts were made to consider Newton's methodology and philosophy in relation to modern physics. Most articles were interesting essays on Newton's work in one specific area and often did not even try to look into the interrelationship between Newton's works. The birthplace of classical empiricism made its mark even in this instance.

Although not having posed the main fundamental questions J. J. Thomson's speech was more profound [in content] than the others. Perhaps the most generalising and fundamental in content was the speech by . . . Dr. E. W. Barnes, the Lord Bishop of Birmingham and a member of the Royal Society who attempted to link Newton's works with the current revolution in modern physics. In his extensive speech[3] the Lord Bishop tried to prove that although "[t]he Thirty Nine Articles of our Church . . . belong to the pre-Copernican period of knowledge. . ." and although "[i]ts theology seems to be associated with crude beliefs as to the history and structure of the universe which were held in antiquity . . . in fact, its theology has been continuously re-shaped by its leading divines, and the process has not yet ended."[4] Theology must find support in and make use of all scientific achievements, both past and present. This is why Newton's name is as precious to the church as it is to science.

The Yorkshire Branch of the Mathematical Association requested a memorial church service for Newton, which concluded the celebrations and was followed by the Bishop's final speech. He, as it were, summed up all the previous speeches. The Bishop noted with satisfaction the unity of science and religion and stressed that the initiative to conclude the celebrations with the religious service belonged to a group of leading scientists. English empiricism has been true to its tradition once more.

In contrast with the English spirit, Einstein's article poses precisely the most general and fundamental questions regarding the influence of Newton's ideas on the character of the development of theoretical physics and on the interrelation of these ideas with the methodology of modern natural science.

Therefore, it is appropriate to say several words about some fundamental concepts proposed by Einstein.

The seventeenth century was a turning point in the history of the development of physics. Started by Galileo, the fight against the scholastic physics of Aristotle had finally succeeded by the beginning of the eighteenth century. The "hidden qualities"—horror vacui[5] were expelled from physics forever. General laws of terres-

---

[2]TN: For J. J. Thomson, see Chap. 3, p. 35, n. 20 and p. 40, n. 37. Sir James Hopwood Jeans (1877–1946) was a British physicist and mathematician. He developed a 'steady state' cosmology and was unusual in his support for philosophical idealism. Sir Horace Lamb (1849–1934) was a British applied mathematician, well-known for his work on hydrodynamics. He wrote a number of influential text books. Sir Richard Tetley Glazebrook (1854–1935) was a leading British physicist.

[3]TN: see Barnes (1927).

[4]TN: Quotations translated by Hessen replaced by original from Barnes (1927).

[5]TN: Latin for "fear of empty space".

trial and celestial mechanics were established. Gassendi and Boyle clearly formulated the principles of atomism.

The main tools for natural research had been created and improved (telescope, microscope and thermometer).

Solid foundations had been laid for starting quantitative and mathematical natural research.

Two sources that feed the reaction against scholastic physics are Bacon's empiricism and the mechanistic principles of Descartes's physics.

Descartes was the most consistent and principal adversary of Aristotle's scholastic physics, i.e. the physics of hidden qualities.

Aristotle called "hidden" those qualities of objects that could not be directly perceived by our senses but could be the cause of the observed actions of objects, e.g. a magnet attracts due to its magnetic force of gravity. This force is the hidden quality, qualitas occulta.

Clearly this methodology could not be used as a tool for scientific research and was forcefully opposed by Descartes.

He stated, "For I openly acknowledge that I know of no kind of material substance other than that which can be divided, shaped and moved in every possible way, and which Geometers call quantity and take as the object of their demonstrations. And [I also acknowledge] that there is absolutely nothing to investigate about this substance except those divisions, shapes, and movements; and that nothing concerning these can be accepted as true unless it is deduced from common notions, whose truth we cannot doubt, with such certainty that it must be considered as a Mathematical demonstration. And because all-natural phenomena can be thus explained, as will appear in what follows; I think that no other principles of Physics should be accepted, or even desired."[6]

This mechanistic principle of interpretation of nature has become both a slogan in the fight against scholasticism and the main methodology of science.

Huygens said, "... true Philosophy, in which one conceives the causes of all natural effects in terms of mechanical motions. This, in my opinion, we must necessarily do, or else renounce all hopes of ever comprehending anything in Physics."[7]

However, the mechanical principle is only one of the components of Descartes's physics.

The method of scientific research was the main question at the time. While Bacon chose experiment as the main starting point, Descartes considered deduction to be the only true approach to cognition. His physics is a wonderful example of the application of rationalism combined with the mechanical principle in the understanding of nature. This is both its advantage and disadvantage. Thanks to this synthesis Descartes succeeded in the creation of a magnificent picture of the world which not only has not lost its significance but is becoming particularly interesting today.

---

[6]TN: Hessen references a Russian translation. The text used here is taken from the English translation in Descartes (1983), Part II, paragraph 64, p. 77.

[7]BH: Huygens, Traité de la Lumière, Paris, 1910, p. 3. TN: English translation used here in Huygens (1969), p. 3.

"The grandeur of Descartes' plan, and the boldness of its execution, stimulated scientific thought to a degree before unparalleled; and it was largely from its ruins that later philosophers constructed those more valid theories which have endured to our own time."[8]

Descartes' physics aspired to give a comprehensive picture of all natural phenomena based on the mechanical view of the world. "In physics . . . after the true origins of material things are found, one moves on to the general question of how the world originated. Then, on to the question of a particular nature of the earth and all other bodies about the earth, e.g. air, water, fire, magnets and other materials. Further on, the nature of individual planets, animals and especially people must be separately studied in order to facilitate the discovery of other useful truths." (Descartes, *Principles of Philosophy*, Letter from the Author to the Translator of this Book (which can serve here as a Preface), translation by Sretensky, p. 9)[9]

Having established the concepts of matter and motion Descartes constructs the entire system of physics following the above plan.

However, his physics is not an encyclopaedic resume of modern knowledge. The science of physics did not exist at the time. Starting with Galileo physics has been going through a period of original accumulation of factual knowledge of nature obtained now not just by observation but also by experiment. This is why Descartes built his physics as a finished rationalistic system, at times contrary to currently known facts (the collisions of the balls).[10]

Newton set himself different tasks in physics.

The main question for Descartes was that of method. Once this method had been found the system of science had to be constructed. While ruthlessly fighting Aristotle's physics of hidden qualities Descartes formulated the main concepts of the mechanistic view of the world as general methodological premises for the study of nature.

---

[8]BH: Whittaker (1910), p. 3.

[9]TN: An English translation of Hessen's loose Russian translation is given. An accurate English translation of the letter is given in Descartes (1983), pp. xvii–xxviii. The relevant portion (pp. xxv–xxvi) is: "The other three parts [i.e. of the Principles] contain everything which is most general in Physics, namely, the explanation of the first laws or Principles of Nature, and the way in which the Heavens, the fixed Stars, the Planets, the Comets, and generally all the universe is composed; then, in particular, the nature of this earth, of air, of water, of fire, and of the loadstone, which are the bodies one can most commonly find everywhere about the earth; and of all the qualities which are observed in these bodies, such as light, heat, weight, and similar things: by which means I believe I have begun to explain all Philosophy in correct order, without having omitted any of those things which ought to precede the last things which I wrote. However, in order to pursue this project to completion, I ought hereafter to explain in the same way the nature of each of the other even more particular bodies which are on the earth, namely, minerals, plants, animals, and, principally, man".

[10]TN: Understanding how Descartes developed his ideas about corpuscular balls (boules in French) as part of his vortex theory requires a complex historical reconstruction. For recent work on this see Schuster (2013), especially Chap. 10 and Appendix 2.

Descartes consistently followed this principle and without hesitation reduced mass to extension (volume).[11] Any other understanding of mass for him was a "hidden quality". But all phenomena had to be explained only on the basis of "shapes and movements".

This uniform and complete construction of the edifice of physics based on the unity of the mechanistic view of the world and the rationalistic method is opposed by Newton's physics as a system of mathematical phenomenology.

"General phenomenology (according to Boltzmann) seeks to describe every group of facts by enumeration and by an account of the natural history of all phenomena that belong to that area, without restriction as to means employed except that it renounces any uniform conception of nature, any mechanical explanation or other rational foundation."[12]

Indeed, although Newton's work is called *Mathematical Principles of Natural Philosophy* we find there no philosophically explained and consistently applied natural and scientific view of the world.

Newton's methodology is a methodology of empiricism dressed in mathematical form.

The conflict between Newtonian and Cartesian physics is a conflict between empiricism and rationalism in [the area of the] study of nature. Notably, Newtonian physics does not succeed as a synthesis of these approaches but as a mathematical phenomenology.

Therefore, Newton's historical contribution and significance can be correctly understood and appreciated only by its comparison with Descartes.

In this respect J. J. Thomson is absolutely right to pay tribute to Descartes in his speech on Newton's work in physics. Interestingly, he points out precisely those general methodological principles which were introduced in physics by Descartes and does not point out that his theories of vortices and the ether were consequences of the above mentioned methodological principles.

Both Descartes's vortex theory and the ether are closely linked to his profound approach to the question of discontinuity and continuity. The issue of long-range and short-range interaction essentially is an issue of discontinuity and continuity. Classical science of the atom accepts atoms and empty space and believes that every action is not an action at a distance but an impulse, an "action from behind" (vis a tergo). But essentially it does not explain anything because a transfer of impulse through a push methodologically is as incomprehensible as an action at a distance.

Maxwell demonstrated in an extraordinarily clever experiment that when one body pushes another, it does not touch it.[13]

---

[11]TN: Descartes's theory of "massiveness" and "solidity" interpreted as the ratio of surface area to volume is discussed in Schuster (2013), Appendix 2.

[12]BH: L. Boltzmann, Entwicklung d. Methoden d. theoretischen Physik. TN: English translation here taken from Boltzmann (1974), p. 95.

[13]BH: Maxwell, *On Action at a Distance*, Russian translation. TN: The original English version is in Niven (1965), LIV, pp. 311–323.

Since Descartes perfectly understood these difficulties, in his vortex and ether theory he wanted to give a synthesis of the atomistic and continuous theory of matter.

The problem of continuity and discontinuity is a cardinal problem of modern physics. This is why the methodological research by Descartes is fresh and interesting today.

Newton approached these problems from the view point of pure phenomenology.

Certainly, Newton posed all these fundamental questions, but his methodological arguments essentially do not comprise an integral system nor serve as a starting point for his constructions.

In this sense one can and should contrast Newton and Descartes, for whom [a] method and main methodological premises make up the foundation and the essence of his physics.

It is rather well known[14] that Newton's empiricism and phenomenology were focused and transformed into a systematic view of the world by Cotes. But it would be totally wrong and against Marxist principles to conclude that there is essentially no contradiction between Newton and Descartes.

We treat Newton's physics as an integral and historically necessary system of physical considerations. It is not Newton's character (which allowed Cotes to make amendments to the *Principia*), that is important to us but the entire system of Newton's ideas because this system resulted in the character and direction of his concrete research.

Any turning point in history should be examined in the fullness of all current conflicting trends.

Luther is incomprehensible and incomplete without the understanding of Müntzer.

The example of the Marxist analysis given by Engels in his *Peasant War* should be applied to our examination of the time of the conflict between Newton and Descartes. In the same way as "An anticipation of communism in fantasy in reality became an anticipation of modern bourgeois conditions",[15] Descartes's physics is a grandiose anticipation of the methodological problems that became significant two centuries later on the basis of the accumulated factual material.

As with any methodology, Newton's physics is an essential stage in the development of exact natural science, but its further development reveals more and more the insufficient and unsatisfactory aspects of its methodological premises.

---

[14]BH: Maxwell wrote: "The doctrine of direct action at a distance cannot claim for its author the discoverer of universal gravitation. It was first asserted by Roger Cotes, in the preface to the Principia, which he edited during Newton's life. According to Cotes, it is by experience that we learn that all bodies gravitate. We do not learn in any other way that they are extended, moveable, or solid. Gravitation, therefore, has as much right to be considered an essential property of matter as extension, mobility or impenetrability.

And when the Newtonian philosophy gained ground in Europe, it was the opinion of Cotes, rather than that of Newton, that became most prevalent . . .". TN: English translation, Niven (1965), p. 316. The idea that action at a distance was the view of Cotes rather than Newton is no longer put forward. However what Newton actually did think is a matter of debate. See, for example https://plato.stanford.edu/entries/newton-philosophy/.

[15]BH: Engels, *The Peasant War in Germany* (in Russian). TN: The English translation is from *The Peasant War in Germany* in Engels (2010), p. 415.

This significance of Newton's physics as a stage of physical research received an excellent evaluation by Maxwell.

". . . it was most essential that Newton's method should be extended to every branch of science to which it was applicable—that we should investigate the forces with which bodies act on each other in the first place, before attempting to explain *how* that force is transmitted. No men could be better fitted to apply themselves exclusively to the first part of the problem, than those who considered the second part quite unnecessary".[16]

Modern natural science owes its independence to its freedom from teleology. It accepts only the causal examination of nature.[17]

One of the battle slogans of the Renaissance was: "true knowledge is knowledge by causes" (vere scire per causas scire).[18]

Bacon emphasized that the teleological view is the most dangerous of the *idola*. The true relations of things are found in mechanical causation. "Nature knows only mechanical causation, to the investigation of which all our efforts should be directed."[19]

A mechanical conception of the universe necessarily leads to a mechanical conception of causation. Descartes laid down the principle of causation (ex nihilo nihil fit) as "an eternal truth."[20]

Mechanical determinism came to be generally accepted on English soil, although it was often interwoven with religious dogma (the "Christian necessarian" sect, to which Priestley belonged). This peculiar combination—so characteristic of English thinkers—is also found in Newton.

The universal acceptance of the principle of mechanical causation as the sole and basic principle for the scientific investigation of nature was brought about by the mighty development of mechanics. Newton's *Principia* is a grandiose application of this principle to our planetary system. "The old teleology has gone to the Devil,"[21] but so far only in the realm of inorganic nature, of terrestrial and celestial mechanics.

The basic idea of the *Principia* consists in the conception of the motion of the planets as a result of the compounding of two forces: one directed towards the sun,

---

[16]TN: English translation Niven (1965), p. 317.

[17]This paragraph and the following six are repeated in "The Social and Economic Roots of Newton's *Principia*", Freudenthal and McLaughlin (2009), p. 67.

[18]TN: Francis Bacon, Novum Organum, Book II, Aphorism II, Bacon (2000), p. 102. (Freudenthal and McLaughlin (2009), p. 67, n. 54).

[19]TN: This quote was not located by Freudenthal and McLaughlin and is presumably mistaken. Bacon did not use the term "mechanical" (Freudenthal and McLaughlin (2009), p. 67, n. 55).

[20]TN: Descartes (1983), Article 49, p. 22.

[21]TN: *Dialectics of Nature*, Engels (1988), p. 475. (Freudenthal and McLaughlin (2009), p. 67, n. 56).

and the other, that of the original impulse.[22] Newton left this original impulse to God but "forbade Him further interference in His solar system."[23]

This unique "division of labour" in the government of the universe between God and causation was characteristic of the way in which the English historians[24] interwove religious dogma with the materialistic principles of mechanical causation, as pointed out by Plekhanov.[25]

The acceptance of the modality of motion, and the rejection of moving matter as *causa sui* was inevitably bound to bring Newton to the conception of the original impulse.[26] From this perspective, the conception of divinity in Newton's system is by no means incidental but is organically connected with his views on matter and motion, as well as with his views on space, in the development of which he was greatly influenced by H. More.[27]

It is at this point that the entire weakness of Newton's general philosophical conception of the universe becomes apparent. The principle of pure mechanical causation leads to the notion of the divine impulse. The "bad infinity" of the universal chain of mechanical determinism ends in the original impulse, thus opening the door to previously rejected teleology.[28]

However, upon having created the world and given an initial impulse to matter, God handed the world over to the rule of mechanical causality. The world, where the law of gravitation applied, continued independently [of God].

In this respect Newton's system was a truly complete system of physical causality, as Einstein stated in his article.

Newton expressed the law of causality in mathematical terms and gave it the appearance still regarded by theoretical physics as the only possible formulation of the causality principle in physics.

The study of the micro-world and internal atomic processes, as well as the accumulation of new experimental data, resulted in the revision of the causality concept within the framework of physical research in the same way as the development of

---

[22]BH: Halley's letter to Newton of 29 June 1686. TN: Turnbull (1960), p. 441. Halley writes of the problem of "making out the Planets motions by a composition of a Descent towards the sun, & an imprest motion".

[23]TN: Freudenthal and McLaughlin (2009), p. 67, n. 57, give the reference "Newton allowed Him the 'first impulse' but forbade Him further interference in his solar system." (*Dialectics of Nature*, Engels (1988), p. 480). They point out that only the second clause of the sentence is in quotes in the Russian.

[24]TN: The same sentence in Freudenthal and McLaughlin (2009) has "philosophers" rather than "historians", presumably a correction.

[25]BH: Plekhanov, *The Role of the Individual in History*, Vol. 8, p. 274 (In Russian). TN: English translation Plekhanov (1976), p. 284. Available via https://www.marxists.org/archive/plekhanov/1898/xx/individual.html, cited 13.03.20.

[26]BH: The initial impulse is the tangential component, of which Engels accused Newton. TN: See n. 23 above.

[27]TN: Henry More. English seventeenth century Cambridge Platonist philosopher.

[28]TN: Engels refers to Hegel's notion of "bad infinity" several times, for example in *Anti-Dühring*, Engels (1988), p. 44. He writes that the "first conclusion drawn from this conception of infinity is that the chain of causes and effects in the world must at some time have had a beginning".

exact natural science resulted in the revolution in the understanding of space and time.

Newton's physics is the physics of the macro-world par excellence. Clearly the transition to studying the micro-world required an innovative approach.

The problem of the law of causality is currently the focus of attention of modern physics; therefore, Einstein poses the question of causality along with the concepts of space and time. The concept of mechanical causality is essentially a space-time construction. Naturally, denial of the causality law poses the question of the possibility of the space-time concept of phenomena. No wonder the causality law question is today's flavour of the day in physics.

The development of Newton's basic mechanics inevitably resulted in the denial of the concept of absolute space and absolute time. Maybe the modern development of physics of atomic processes faces the necessity of giving up the causality law?

Einstein finishes his article by pointing out that it is unlikely today to find someone who would dare to say whether the principle of causality should be abandoned. However, we shall attempt to examine this question from the view point of dialectical materialism.

The question of chance and necessity, commonly known in physics as statistical and dynamical laws, was posed at the time of the development of the kinetic theory of matter. Maxwell was the first to pose this question in its general principal form. As Planck justly noted the dualism between statistical and dynamical laws is closely linked to the dualism between the macro-world and the micro-world.

Newton's physical laws are dynamical in character precisely because they are laws of the macro-world.

A law is usually called dynamical if it describes phenomena where the state of a system at a given moment in time defines both its future and its past state. If a planet's position and velocity are given at a given moment in time, then this clearly defines the planet's behaviour both in the past and the future.

There is no place for probability in Newton's mechanics. Each future state is unequivocally defined by its previous one. This is why Einstein called Newton's system one of complete physical causality. It is for this reason that he expressed Newton's laws in the form of differential equations, i.e. in the form of the relation between infinitely small elements of values that are used in the equations. Accordingly, the law of continuous causality is expressed in a mathematical form, because a given state of a system at an infinitely small element of time defines its next state, and the transition between the states occurs continuously. In this way the need for the causal understanding of nature received its preliminary completion by Newton.

When physics steps aside from the purely phenomenological viewpoint and penetrates the depth of micro-world phenomena the old methods become inadequate.

Primarily, a dynamical examination of phenomena becomes insufficient.

Indeed, a study of phenomena based on molecular processes becomes a study of collective behaviour rather than the study of the behaviour of individuals.

Therefore, the concepts of chance and probability start playing a prominent part in the latest physics.

If planets in a solar system are at a certain given position then their next position necessarily follows from it. If two bodies with different temperatures are examined, then the transition of heat from a warmer to a cooler one is most probable, however, the reverse transition from a cooler body to a warmer one is possible although is immeasurably less probable.

The research of thermodynamic phenomena from a purely phenomenological viewpoint resulted in establishing an impassable gap between reversible and irreversible phenomena. Only by introducing the concept of a statistical law Boltzmann managed to breach this gap and to show the relativity of the concept of irreversibility. However, at the same time the question arose concerning the very essence of dynamical and statistical laws. Which of the two should be accepted as the main one and, as it were, the absolute one? Should one try to reduce any statistical law to a dynamical one or are they equal methods of phenomenological research? These questions are also relevant today, because research of atomic processes is one of the main tasks of modern physics.

We have noted above the connection between dynamical laws and phenomenology.

According to Born, the difference in approach to the study of microcosmic phenomena in classical and in modern physics is, "[t]he classical theory introduces the microscopic coordinates which determine the individual process, only to eliminate them because of ignorance by averaging over their values; whereas the new theory gets the same results without introducing them at all."[29]

Born further reasoned, that we did not possess means for the observation of the behaviour of each individual atom or electron during a complicated experiment.

At best, we can observe a final and an original state. We know neither what happens between the two, nor how an electron behaves during this time. Original and final states are not connected by a definite causal chain of states as happens in the dynamical laws of Newton's classical physics. Therefore, an original state does not define the final state absolutely but only probably. If we know the position of the earth, we can definitely and precisely define the position which it will take in a certain period of time. However, if we know the state of a given aggregate of atoms we can define its next state only with a certain measure of probability.

Furthermore, some researchers also question determinism for the following reasons: with the transition from classical physics to quantum theory we enter the realm of discontinuous processes. "Physical quantities are *not* continuously propagated through space; physical motions are not invariably continuous; . . . What remains of determinism is not necessarily more than statistical. If we work with a great many similar atoms, or repeat very often experiments with a few, then we always get a result in agreement with the principle of determinism." However, these results significantly differ in character from the results provided by classical physics. "The classical calculation gives us information about our specific system of planets. The

---

[29]BH: Born (1927a, b). TN: The quotation is taken from the first reference. It is also available in Born (1969).

quantum theoretical calculation does not, in general, tell us anything about a single atom, but only about the mean properties of an assembly of similar atoms."[30]

Thus, the statistical laws of quantum processes result from their discontinuous character.

Classical physics has mainly researched a sequence of separate states, i.e. the very performance of a process. Conversely, as noted by Born, in new quantum mechanics "the question of the course of phenomena had practically disappeared"[31] from the sphere of research.

The laws of the new mechanics are mainly statistical laws. However, since we do not examine the sequence of events but only the final empirically observed states, would it be prudent to say that the atomic processes are unequivocally defined for their duration? Can we talk about the causal study of phenomena if a final state could only probably be determined from an initial state? Does not this concept of causal phenomena pose a general question on the causal sequence of events?

These are the questions raised by the new quantum mechanics and this is what Einstein means when he says that the law of causality refuses to work when confronted with the difficulties brought about by the most recent achievements in physics.[32]

As we have seen in modern physics, this renouncement of the causal study of phenomena is understood as a renouncement of the establishment of a continuous connection between initial and final states and its replacement by the determination of the probability of a given state.

As we have seen earlier, while modern quantum mechanics, unlike the classical case, does not even introduce micro-world coordinates and is limited to empirically observed values, it is clear that "[it] provides no means for the determination of the position of particles in space and time".[33]

A certain statistical phenomenology has replaced dynamical description. It is in this sense that the time-space description is renounced.

The answer to the question whether the modern development of physics leads to the renouncement of the law of causality and of space-time description can be found only if the question of the relation between statistical and dynamical laws, i.e. the question of necessity and chance, is posed correctly.

The law of causality may be renounced only if necessity and chance are opposed metaphysically.

Indeed, if dynamical laws are considered to be the only expression of physical causality and have no place for chance, and if dynamical laws are opposed by

---

30BH: See Jordan (1927a). TN: The quotation is taken from the original English.

31TN: Born (1927a). Quotation from original. The word "trajectory" would perhaps be better than "course" which is used in the reference.

32BH: Lately the problem of causality and statistical laws has become most relevant particularly after the Schrödinger theory, which seemingly returned to the dynamical concept of molecular phenomena, received a purely statistical explanation. See P. Jordan for an essay on Schrödinger's work, *Naturwissenschaften*, 5 May 1927. TN: i.e. Jordan (1927b).

33BH: P. Jordan, loc. cit. TN: This reference to Jordan (1927b) is a mistake. It should be Born (1927a). Quotation from original English.

statistical laws which are based only on the concept of probability, then it is reasonable to consider such [statistical] laws as an antithesis to causality and complete determinism.

It was Engels who first pointed out the significance of the concept of chance in theoretical natural science. He explained that the metaphysical opposition of chance and necessity could not suffice for the development of research and that naked mechanical determinism could not serve as an adequate research tool. Chance is an objective category. "Common sense, and with it the majority of natural scientists, treats necessity and chance as determinations that exclude each other once for all. . .

And then it is declared that the necessary is the sole thing of scientific interest and that the accidental is a matter of indifference to science. . .

That is to say: what can be brought under general laws is regarded as necessary, and what cannot be so brought as accidental. . .

In opposition to this view [the accidental] there is determinism, which passed from French materialism into natural science, and which tries to dispose of chance by denying it altogether. According to this conception only simple, direct necessity prevails in nature. . .

With this kind of necessity, we likewise do not get away from the theological conception of nature. . .

Hence chance is not here explained by necessity, but rather necessity is degraded to the production of what is merely accidental. If the fact that a particular pea-pod contains six peas, and not five or seven, is of the same order as the law of motion of the solar system, or the law of the transformation of energy, then as a matter of fact chance is not elevated into necessity, but rather necessity degraded into chance. . .

In contrast to both conceptions, Hegel came forward with the hitherto quite unheard-of propositions that the accidental has a cause because it is accidental, and just as much also has no cause because it is accidental; that the accidental is necessary, that necessity determines itself as chance, and, on the other hand, this chance is rather absolute necessity.[34] Natural science has simply ignored these propositions as paradoxical trifling, as self-contradictory nonsense, and, as regards theory, has persisted on the one hand in the barrenness of thought of Wolffian metaphysics, according to which a thing is either accidental or necessary, but not both at once; or, on the other hand, in the hardly less thoughtless mechanical determinism which in words denies chance in general only to recognise it in practice in each particular case."[35]

However, to renounce fatalistic determinism does not mean to renounce the law of causality.

---

[34]BH: Logik, II, Book III, 2: 'Die Wirklichkeit'. TN: Hegel (2010), Book Two, Section III, Actuality, pp. 465–505.

[35]BH: Engels, Archive II, 193–195 (in Russian). TN: *Dialectics of Nature*, Engels (1988), pp. 498–501. (The sentences quoted are taken from over four pages. The dots indicate that they do not follow directly from one another.)

To accept chance as a real and objective category and not a simple consequence of our ignorance of causal chains certainly does not mean to equate chance and lack of causality and to introduce lack of causality as a sort of additional "postulate."[36]

Such an understanding of chance results in a different approach to the problem of statistical laws.

Indeed, statistical laws are based on the concept of probability of phenomena; the concept of probability is based on the concept of chance.

Therefore, if we, on the one hand, give up the fatalistic concept of determinism, and [on the other hand,] accept chance not just as a consequence of our ignorance but an objective category, then contradiction between dynamical and statistical laws disappears. They do not exclude but imply each other. They are both legitimate and necessary. Statistical laws are not the consequence of our insufficient knowledge of processes but an objectively necessary method of research rooted in the peculiarities of the studied phenomena. Engels's concept of chance and necessity gives us a clue [to the solution] not through renouncement of causality but through the correct synthesis of necessity and chance, and of dynamical and statistical laws. Since chance is not our unknown necessity and is an objective and not a subjective category, then a particular science must decide what is chance for a given process, and therefore, also resolve the question which law is most appropriate for the study of a given group of phenomena, i.e. statistical or dynamical. And these two laws cannot be regarded as mutually exclusive. If an initial [and] final state in a statistical law are not linked by a continuous sequence of states, this only signifies that for a given phenomenon at a given stage of a certain study a number of intermediate states are accidental. Whereas the final state is always a necessary consequence of elementary processes—the components of this state, these elementary processes are accidental in this chain of events with regard to the entire process.

Engels's methodological views were confirmed in a very interesting work "On the concept of chance and on the origin of probability laws in physics" by Smoluchowski.[37]

Smoluchowski considers the accepted concepts in physics of chance and probability utterly unsatisfactory.

He says: "My main thought is that the *objective* aspect of the concept of probability hitherto totally neglected must be put in the right light."[38]

---

[36]BH: See Deborin, "Our Disagreements", in *Letopisi marksizma* (*Chronicles of Marxism*), Vol. 2, 1926, pp. 9–11 (in Russian).

[37]BH: Smoluchowski (1918). TN: The following quotations given by Hessen are only rough translations. Therefore we have translated from Hessen's Russian text into English and given the original German with page numbers in footnotes.

[38]TN: "Im übrigen bezweckt dieselbe selbstverständlich keineswegs eine allseitige und endgültige Aufklärung des ganzen damit zusammenhängenden Komplexes philosophischer Fragen, sondern will nur eine Anregung zu weiteren Untersuchungen in einer bestimmten Richtung geben, indem einige *Leitgedanken* hervorgehoben werden, welche die bisher allzusehr vernachlässigte *objektive Seite des Wahrscheinlichkeitsbegriffes ins rechte Licht setzen* sollen." (Italics in original, p. 253).

"... Probability laws cover those events whose occurrence depends on chance."[39]

"If one views chance, as popular theories do, as the negation of ... laws, then these contradictions are indeed completely unbridgeable."[40]

"Thus, insofar as [their] application in theoretical physics is to be considered, all probability theories which perceive chance as an unknown 'partial cause' must ... from the outset be regarded as insufficient."[41]

"The physical probability of an event can only depend on the conditions which influence its coming about, but not on the degree of our knowledge!"[42]

However, unlike in mechanical determinism, if one looks at an objective aspect of probability and chance and stops considering chance an unknown necessity, then one also has to change one's view on the significance of statistical laws. This explains why Smoluchowski, upon establishing the flaw in the old concepts of chance and probability, has a different approach to the problem of statistical laws.

He says, "I am well aware that this concept of chance stands in contrast to the generally accepted understanding, whereby its essential aspect is the partial ignorance of causes. So may the following be said as attestation of our assertion: probability theory's application in kinetic gas theory would retain its legitimacy even if we knew the exact constitution of molecules and their initial positions, and were able to follow their motion with mathematical precision for all times."[43]

Thereby, statistical regularity and statistical laws, i.e. laws based on the concept of chance, are not a simple result of our lack of knowledge but constitute an equally legitimate method of natural research. At the same time the general concept of determinism is preserved. But this determinism is not the limited mechanical determinism. Determinism and statistical laws, as well as necessity and chance should not be considered to be mutually exclusive. And Smoluchowski agrees with this.

He says, "I think that it would also be an extremely important result for a philosopher if it can be demonstrated, albeit in a narrow field of physics, that the concept of probability possesses a strictly objective meaning in its usual interpretation as a

[39]TN: Die Frage, welche Ereignisse in den Geltungsbereich der Wahrscheinlichkeitsrechnung fallen, wird wohl allgemein dahin beantwortet: diejenigen, *deren Eintritt vom Zufall abhängt.* (Italics in original, p. 253).

[40]TN: Betrachtet man in populärer Weise den Zufall als die Negation des Gesetzmäßigen, so sind diese Widersprüche gewiß vollständig unüberbrückbar, (pp. 253–254).

[41]TN: Offenbar sind also, soweit die Anwendung in der theoretischen Physik in Betracht kommt, *alle Wahrscheinlichkeitstheorien* von vornherein als *ungenügend* zu betrachten, *welche den Zufall als "unbekannte Teilursache" auffassen*, (Italics in original, p. 254).

[42]TN: *Die physikalische Wahrscheinlichkeit eines Ereignisses kann nur von den Bedingungen abhängen, welche sein Zustandekommen beeinflussen, aber nicht von dem Grade unseres Wissens!* (Italics in original, p. 254).

[43]TN: Ich bin mir wohl bewußt, daß dies im Gegensatz zu der allgemein üblichen Auffassung steht, welche eine teilweise Unkenntnis der Ursachen als das wesentliche hier in Betracht kommende Moment ansieht, darum sei als Beleg für unsere Behauptung bemerkt: Die Wahrscheinlichkeitsrechnungen der kinetischen Gastheorie würden ihre Berechtigung auch dann behalten, wenn wir die Beschaffenheit der Moleküle, deren Anfangslagen usw. absolut genau kennen würden und imstande wären, deren Bewegungen mathematisch exakt für alle Zeiten zu verfolgen (p. 254).

frequency law of accidental events, and that the concept and the genesis of chance can be determined with precision while remaining firmly rooted in determinism."[44]

We have by no means exhausted the problem of the character of a physical law in these short notes. But in any case we can confidently state that the achievements of modern physics give no reason to doubt the causal connection of phenomena and the space-time concept. The development of our knowledge has resulted in the necessary change of Newton's concept of absolute space and time. Similarly, the development of quantum theory has posed the question of insufficiency of the concept of continuous mechanical determinism. Essentially, the question of the renouncement of determinism is just evidence for the insufficiency of the metaphysical concept of causality.

Back in 1873, at the time of birth of the kinetic gas theory, with his ingenious insight Maxwell [45] asked whether modern developments of physics had any arguments against determinism.

He answered as follows, "If, therefore, the cultivators of physical science from whom the intelligent public deduce their conception of the physicist . . . are led in pursuit of the arcane of science to the study of the singularities and instabilities, rather than the continuities and stabilities of things, the promotion of natural knowledge may tend to remove that prejudice in favour of determinism which seems to arise from assuming that the physical science of the future is a mere magnified image of that of the past."[46]

The development of science is not a simple quantitative accumulation of factual data. It is inextricably connected with the development and change of the main methodological concepts. Science in its development inevitably grows out of the constraints of old notions and concepts. This growth must be accompanied by different idealistic vacillations which in most cases signal difficulties that are unresolvable within the framework of the old view of the world.

We witness a comparable situation in the question of the role and significance of causality in modern physics.

---

[44]TN: Es scheint uns aber ein auch *für den Philosophen äußerst wichtiges Ergebnis* zu sein, wenn sich auch nur auf einem beschränkten Gebiet – dem der mathematischen Physik – zeigen läßt, daß der *Begriff der Wahrscheinlichkeit*, in der üblichen Bedeutung eines *gesetzmäßigen Häufigkeitswertes zufälliger Ereignisse, eine streng objektive Bedeutung besitzt*, daß man den *Begriff und die Genese des Zufalls* genau präzisieren kann, auch wenn man am Determinismus festhält, . . . (Italics in original, p. 262).

[45]BH: Maxwell looks into the problem of necessity and chance applied to physics in a small article published by Campbell and Garnett–Maxwell's paper was given to a philosophical group at Cambridge (Club of Seniors). The article is titled "Does the progress of Physical Science tend to give any advantage to the opinion of Necessity (or Determinism) over that of the Contingency of Events and the Freedom of the Will?"TN: See Campbell and Garnett (1882), pp. 209–213. Hessen mistakenly wrote that the talk was given to the Cambridge Philosophical Society "Club of Seniors". According to Campbell and Garnett it was actually given on 11th February 1873 to the Eranus Club, a small group of Cambridge University Seniors–Campbell and Garnett (1882), pp. 179 and 209.

[46]TN: Original English taken from Campbell and Garnett (1882), p. 213.

However, one thing is obvious to us: "It is mainly because the physicists did not know dialectics that the new physics strayed into idealism. The basic materialist spirit of physics, as of all modern science, will overcome all crises, but only by the indispensable replacement of metaphysical materialism by dialectical materialism."[47]

We have attempted to demonstrate that the substitution of the metaphysical opposition of necessity and chance by the dialectical concept of causality, pointed out by Engels, indeed provides a solution to the crisis that has led to the renouncement of causal and space-time concepts of phenomena.

References

Bacon, F. (2000). *The new organon.*, Trans. Lisa Jardine and Michael Silverthorne Cambridge: Cambridge University Press.
Barnes, E. W. (1927). The bicentenary of Newton's death. *Supplement to Nature, 119*(2995), 21–24.
Boltzmann, L. (1974). *Theoretical physics and philosophical problems.* Dordrecht: Reidel.
Born, M. (1927a). Physical aspects of quantum mechanics. *Nature, 119*(2992), 354–357.
Born, M. (1927b). Quantenmechanik und Statistik. *Die Naturwissenschaften (The Science of Nature), 15*(10), 238–242.
Born, M. (1969). *Physics in my generation.* New York: Springer.
Campbell, L., & Garnett, W. (1882). *The life of James Clerk Maxwell with the selection of his correspondence and occasional writings.* London: Macmillan.
Descartes, R. (1983). *Principles of Philosophy (trans. Miller, V.R and Miller, R.P).* Reidel, Dordrecht.
Einstein, A. (1927). Newton's Mechanik und ihr Einfluß auf die Gestaltung der theoretischen Physik. *Die Naturwissenschaften (The Science of Nature), 15*(12), 273–276.
Engels, F. (1988). *Marx engels collected works* (Vol. 25). Moscow: Progress Publishers.
Engels, F. (2010). *Marx engels collected works* (Vol. 10). London: Lawrence and Wishart.
Freudenthal, G., & McLaughlin, P., (ed.) (2009). *The social and economic roots of the scientific revolution: Texts by Boris Hessen and Henryk Grossmann*, Boston Studies in the Philosophy of Science, p. 278. Springer, Berlin.
Hegel, G. W. F. (2010). *Science of logic. trans. G. di Giovanni.* Cambridge: Cambridge University Press.
Huygens, C. (1969). *Treatise on light.* New York: Dover.
Jordan, P. (1927a). Philosophical foundations of quantum theory. *Nature, 119*(2998), 566–569.
Jordan, P. (1927b). [Review of] Schrödinger,E., Abhandlungen zur Wellen-mechanik. *Die Naturwissenschaften (The Science of Nature), 15*(18), 412–413.
Lamb, H. (1927). Newton's work in mechanics. *Supplement to Nature, 119*(2995), 33–35.
Lenin, V. (1977). *Collected works* (Vol. 14). Moscow: Progress Publishers.
Niven, W. D. (Ed.). (1965). *The scientific papers of James Clerk Maxwell,* (Vol. II). New York: Dover.
Plekhanov, G. V. (1976). *Selected philosophical works II.* Moscow: Progress Publishers.
Schuster, J. (2013). *Descartes-agonistes: Physico-mathematics, method & corpuscular-mechanism* (pp. 1618–1633). Heidelberg: Springer.
Thomson, J. J. (1927). Newton's work in physics. *Supplement to Nature, 119*(2995), 36–40.

[47]BH: Lenin, Materialism and Empirio-criticism, p. 312, 1920 edition (in Russian). TN: English translation taken from Lenin (1977). Although Hessen gives one page number, in fact the two sentences are from pp. 262 and 306 in the English edition.

Turnbull, H. W. (Ed.). (1960). *The correspondence of Isaac Newton II*. New York: Cambridge University Press.
Smoluchowski, M. (1918). Über den Begriff des Zufalls und den Ursprung der Wahrscheinlichkeitsgesetze in der Physik. (On the Concept of Chance and the Origin of the Laws of Probability in Physics). *Die Naturwissenschaften (The Science of Nature)*, *6*(17), 253–263.
Whittaker, E. (1910). *A history of the theories of aether and electricity*. Longmans, Green and Co., London.

Chapter 5
Marian Smoluchowski (On the Tenth Anniversary of His Death)

Boris Hessen

In September of this year (1927) ten years will have passed since the death of Marian Smoluchowski[1] His works are of outstanding general value not solely for physicists, but their methodological value is also very significant.

Atomic science flourished in the second half of the nineteenth century due to works by Clausius, Maxwell and Boltzmann but began to fall from grace among physicists by the end of the nineteenth century. The reality of atom was questioned while the attraction of "overcoming the materialism of natural science" increased.

In the introduction to his classical work on kinetic gas theory Boltzmann wrote in 1898 with regret, "it would be a great tragedy for science if the theory of gases were temporarily thrown into oblivion because of a momentary hostile attitude toward it, as was for example the wave theory because of Newton's authority."[2] Smoluchowski's work on the theory of Brownian motion presented a new brilliant proof of the reality of atoms. Einstein noted that the universal acceptance of kinetic theory, mainly due to Smoluchowski's work, is dated to this time, as well as the confidence of physicists in the reality of atoms.

---

[1] BH: A general evaluation of Smoluchowski's work was given by Einstein (1917) and Sommerfeld (1917).

[2] TN: English translation from Boltzmann (1995, p. 192).

TN: Translated from *Pod Znamenem Marksizma* (*Under the Banner of Marxism*), 1927, No. 9, pp. 144–148. The preface to a Russian translation of "On the concept of chance and on the origin of probability laws in physics", von Smoluchowski (1918).

B. Hessen (✉)
Abingdon, UK
e-mail: chris.talbot@protonmail.com

C. Talbot and O. Pattison (eds.), *Boris Hessen: Physics and Philosophy in the Soviet Union, 1927–1931*, History of Physics,
https://doi.org/10.1007/978-3-030-70045-4_5

However, by no means is the significance of Smoluchowski's work limited by this. Boltzmann in his work destroyed the metaphysical gap between reversible and irreversible processes. He demonstrated that the "global clock does not need winding up".[3]

Thanks to Smoluchowski's work Boltzmann's concept received a brilliant experimental proof and a final theoretical completion. The huge methodological significance of eliminating the metaphysical difference between reversible and irreversible processes is really obvious. If we take Clausius's view, then as Engels wrote with brilliant insight, "Clausius' second law, etc., however it may be formulated, shows energy as lost, qualitatively if not quantitatively. *Entropy cannot be destroyed by natural means but it can certainly be created.* The world clock has to be wound up, then it goes on running until it arrives at a state of equilibrium from which only a miracle can set it going again. The energy expended in winding has disappeared, at least qualitatively, and can only be restored by an *impulse from outside*. Hence, an impulse from outside was necessary at the beginning also, hence, the quantity of motion, or energy, existing in the universe was not always the same, hence, energy must have been created, i.e., it must be creatable, and therefore destructible."[4]

Boltzmann succeeded in eliminating this flaw and interpreting the natural processes with the help of dialectics because he used a statistical approach to the molecular processes. What was considered irreversible in the past in Boltzmann's view was fundamentally reversible, but the *probability* of reversing the processes that are considered practically irreversible, is vanishingly small (but not equal to zero!).

If we put a pan with water on a primus the heat transfers from the flame to the water and the water boils. This is something routinely observed in everyday life: heat transfers from a body with higher temperature to a body with the lower one. As we have never in our human experience observed the reverse, we are convinced that there are fundamentally irreversible processes, e.g. the transfer of heat from a warmer body to the less warm one.

However, if heat is nothing more than the motion of molecules, then it is not clear at all why an aggregate of molecules where each one performs a motion that is fundamentally reversible, results in such an irreversible process as the transfer of heat from a warmer body to the less warm one.

Boltzmann's contribution is that on the basis of the kinetic theory he introduced a concept of the *probability* of a process continuing in a certain direction, instead of the impossibility of reversing the process. If we put a pan on a hob, then the probability of the water in the pan boiling is so high that in practical terms we assume it to be the case. However, it is quite possible that the water in the pan freezes, i.e. the heat from the pan transfers to the hob; this is not impossible but very improbable.

In this case, as after the revolution performed by Copernicus, nothing changes in our practical life, but our theoretical views turn completely upside down.

[3]TN: Reference to Engels on Clausius—see quotation in next paragraph.

[4]TN: English translation from *Dialectics of Nature*, Engels (1988, p. 563). Engels adds "*Ad absurdum*!" Italics as in the original and as given by Hessen.

This turn could be achieved only thanks to the development of the kinetic theory of matter. However, the kinetic theory of matter regards a body as an aggregate of a vast number of atoms. This theory made physicists widely use the methods that are most appropriate for the study of collectives, i.e. statistical methods. Since all natural phenomena are based on atomic and internal atomic processes, statistical methods become more and more one of the most important tools in physics. Boltzmann's brilliant results are closely connected to the vital importance of statistical methods in his work.

As justly noted by Sommerfeld, Smoluchowski is a direct successor and continuator of Boltzmann's approach. "Statistics was as vital for him as air".[5]

Lately the statistical approach has become more and more acceptable and popular in physics. Indeed, it became essential to each physicist "like air".

However, while a dynamical concept of natural laws is methodologically easier and clearer, a statistical one poses a whole number of deep methodological questions, first and foremost the problem of causality and chance.[6]

The statistical approach requires a deeper development of causal laws. Probability theory is the mathematical apparatus used by the statistical approach. Therefore, research into the methodological foundations of the statistical method necessarily results in the research into the foundations of the theory of probability. But the concept of probability is closely connected with the concept of chance.[7] This is why the wide popularity of the statistical method highlights the problem of causality, necessity and chance.

Classical physics' interpretation of these concepts becomes inadequate.

Lack of clarity and confusion around these basic concepts leads to the rejection of the law of causality and the resurrection of teleological views etc.

What is the essence of chance? What is the significance and where are the boundaries of applicability of the statistical method in physics? These are the unavoidable questions posed by modern physics.

We all recently witnessed fierce attacks on Dialecticians[8] who dared to suggest that chance was not a subjective category, i.e. a consequence of our ignorance, but a real *objective* category.

Clearly the answer to this question is of immense importance, including for physics. Indeed, if chance is a result of the limitation of our knowledge then the statistical method acquires a subjective colouring. It becomes a temporary crutch for our ignorance. It is impossible to define an objective criterion for the conditions and boundaries of its application. All such criteria will have a subjective colouring similar to our ignorance. Smoluchowski's article below is particularly significant for

---

[5]BH: Sommerfeld (1917, p. 537).

[6]BH: On statistical and dynamical concepts related to the problem of causality in modern physics see: On the bicentenary of Isaac Newton's death. Foreword to articles by A. Einstein and J. J. Thomson by Boris Hessen, *Pod Znamenem Marksizma* (*Under the Banner of Marxism*), 1927, 4, pp. 152–165. TN: See Chap. 4.

[7]TN: Here and in the following "chance" is taken as the translation of the Russian sluchainost'.

[8]TN: i.e. on the Deborinite philosophers by the Mechanists.

us as it both fully confirms Hegel's and Engels's views on chance as an objective category and defended by Dialecticians and concretises this concept using physical examples.

This article is the last work by Smoluchowski. It was published after his death. It analyses the main ideas that are vital for the understanding and evaluation of the statistical method. This is the only purely methodological work by Smoluchowski.[9] The choice of its topic signifies the utmost importance the author attributed to the analysis of certain ideas.

Smoluchowski highlights the *objective* side of the concepts of probability and chance as his main thought in the article.

He says, "…all probability theories which perceive chance as an unknown 'partial cause' must …from the outset be regarded as insufficient. The physical probability of an event can only depend on the conditions which influence its coming about, but not on the degree of our knowledge."[10]

But if chance is an objective category then its essence requires an objective definition and it is necessary to show the conditions when probability theory, i.e. the statistical research method, can be applied and must be applied. Smoluchowski's article looks into these questions. A methodological analysis of the concept of chance is given by a detailed study of simple cases that, as it were, serve as a "model of chance events."

Further, the usual interpretation of chance opposes the concepts of chance and necessity: an event is either necessary or accidental. One excludes the other. Chance is the antithesis of the necessary and the regular. But once we adopt this metaphysical opposition of the accidental and the necessary, i.e. regular, then we unavoidably arrive at the contradiction which Smoluchowski formulated as follows. If we adopt the viewpoint of absolute metaphysical determinism then how can chance arise at all? How can regular causes lead to chance events? If we try to resolve this question by declaring chance a subjective category and a consequence of our partial ignorance then another difficulty immediately arises: objectively there is no chance. Everything that is happening is strictly and singularly determined. However, in our practice and in science we calculate the results of chance (even as a subjective category). The work of an insurance company is a suitable example. So how can one calculate the results of chance? How can *accidental* causes result in *regular* actions? Although we abstractly suppose that chance exists as an unknown necessity, in each specific case we do not know this necessary connection and even do not attempt to establish it. However, the result of the calculated chances produces a steady regularity. "If

[9]BH: Smoluchowski made a brilliant review of his own works on physics in von Smoluchowski (1913, 1914, 1916).

[10]TN: The same quotation is used in Chap. 4, p. 56, n. 41. German original: …*alle Wahrscheinlichkeitstheorien* von vornherein als *ungenügend* zu betrachten, *welche den Zufall als "unbekannte Teilursache" auffassen. Die physikalische Wahrscheinlichkeit eines Ereignisses kann nur von den Bedingungen abhängen, welche sein Zustandekommen beeinflussen, aber nicht von dem Grade unseres Wissens!* (Italics in original, p. 254).

one views chance, as popular theories do, as the negation of ...laws, then these contradictions are indeed completely unbridgeable."[11] states Smoluchowski.

Yet this contradiction must be resolved, and Smoluchowski shows how such contradictions are resolved if one rejects the metaphysical opposition of chance and necessity (regularity) and accepts that chance is an objective category.

In his article *The Role of the Individual in History*[12] Plekhanov gave brilliant examples of concretisation of the dialectical concept of chance applied to social processes based on the acceptance of chance as an objective category and on the dialectical synthesis of the concepts of chance and necessity. Smoluchowski's article specifies the dialectical concept of chance applied to physical phenomena.

This is why this article is of special interest to Marxists and presents an obvious proof of fruitfulness of the dialectical concept of chance.

Editor's Note—CT

As statistical laws became increasingly important at the micro-level, with many quantum physicists arguing that causality could be abandoned entirely, it was necessary to make an assessment from the standpoint of Marxist philosophy. Hessen clearly concentrated on statistical physics, as is clear from his research work with Mandelstam.[13] Hessen based himself on Engels (who derives his ideas from Hegel) in a section of *Dialectics of Nature*. Engels called for chance to be taken as an objective category, with a dialectical relationship between causality and chance.[14] Hessen considers chance in the last part of Chap. 4, here in Chap. 5 as well as the exposition in Chap. 8.

Hessen's emphasis on the only philosophical paper by the Polish theoretical physicist Marian Smoluchowski, published posthumously in 1918, deserves special attention. While Smoluchowski is well known in the history of statistical physics as a major figure—he can be said to have originated the whole subject of stochastic processes—there is practically no mention of his 1918 paper. In an introduction to a collection of some of Smoluchowski's technical papers translated into English,[15] a leading mathematician in this field, Mark Kac, mentions it briefly, stating that Smoluchowski's "claims are modest" and "the article is full of sharp and incisive observations and

[11]TN: This quotation is also used in Chap. 4, p. 46, n. 40. German original: Betrachtet man in populärer Weise den Zufall als die Negation des Gesetzmäßigen, so sind diese Widersprüche gewiß vollständig unüberbrückbar (pp. 253–4).

[12]TN: Plekhanov (1976). Available via https://www.marxists.org/archive/plekhanov/1898/xx/individual.html, cited 13.03.20.

[13]To give some indication of Hessen's interests, note that he published a paper "The Interpretation of the Ergodic Hypothesis by the Theory of Probability", published in Uspekhi Fizicheskikh Nauk (Advances in Physical Sciences), No. 5, 1929, pp. 600–629. We considered translating it but realised that it was an entirely technical introduction to ergodic theory in statistical mechanics as understood at that time, hardly suitable for this collection.

[14]See Chap. 4, p. 54, notes 34–35, Chap. 8, p. 104, n. 10, Chap. 11, pp. 150–151, n. 26.

[15]Ingarden (1999) (Containing a brief biography by Smoluchowski's son, as well as introductions by Kac and the astrophysicist Subrahmanyan Chandrasekhar, this seems to be the only book specifically on Smoluchowski in English).

it leaves no doubt that the author has given the subject much time and thought."[16] The article is no doubt "modest"—in line with Smoluchowski's character—but we suggest that Hessen was correct in giving it some attention. Though no doubt Smoluchowski did not realise it, it does gives powerful support to the dialectical materialist view of causality and chance. Sections of Hessen's translation of the article are translated here into English with the corresponding German in footnotes.[17]

A brief but useful exposition of Smoluchowski's 1918 paper is given by von Plato in his well-known history of probability.[18] Von Plato includes an English translation of a paragraph of the paper,[19] a longer version of one of the translations given here.[20] He points out that Smoluchowski's approach to chance is an objective one[21]:

> First, chance is defined as instability, the typical element in many games of chance. Second, it is required that a physical and objective notion of probability be determined, not from our degree of ignorance concerning an event, but from the conditions that have an effect on its occurrence

Von Plato also notes Smoluchowski's idea that a small variation in a cause can give rise to a great variation in effect, which was, of course, taken up in modern chaos theory. By considering a simple mathematical model Smoluchowski explains that "It shows that the *apparent contradiction* [between chance and lawlike effects of causes] *does not exist* and that chance—in the sense of physics—can very well be brought by exactly defined lawlike causes."[22] In Hessen's terminology there is no "metaphysical opposition of chance and necessity".[23]

Finally it is worth adding that the dialectical conception of causality and chance, based on *Dialectics of Nature*, was the viewpoint of David Bohm in his approach to quantum theory as set out in *Causality and Chance in Modern Physics*[24] with no knowledge of Smuluchowski's 1918 paper. It enabled Bohm to challenge the viewpoint of absolute indeterminism or randomness that is central to "standard" quantum mechanics.

References

Bohm, D. (1957). *Causality and Chance in Modern Physics* (p. 1984). Routledge and Kegan Paul: London, first edition. Second edition with new preface by Bohm.

[16] Ibid., p. 20.

[17] Chapter 4, p. 55, n. 38, p. 56, notes 39–43 and p. 57, n. 44; Chap. 5, p. 64, n.10 and p. 65, n. 11; Chap. 8, p. 106, notes 16–18.

[18] von Plato (1994), pp. 171–173.

[19] von Smoluchowski (1918, p. 262).

[20] Chapter 4, p. 57, n. 44.

[21] von Plato (1994, p. 171), referring to von Smoluchowski (1918, p. 254).

[22] von Plato (1994, p. 173) quoting from von Smoluchowski (1918, p. 262), italics in original.

[23] See p. 64 above.

[24] Bohm (1957).

Boltzmann, L. (1995). *Lectures on gas theory* (S. G. Brush, Trans.). New York: Dover. Reprint of 1964, University of California edition.
Einstein, A. (1917). *Marian v Smoluchowski. Die Naturwissenschaften* [The Science of Nature], *5*(50), 737–738.
Engels, F. (1988). *Marx Engels collected works* (Vol. 25). Moscow: Progress Publishers.
Ingarden, R. S. (Ed.). (1999). *Marian Smoluchowski: His life and scientific work.* Warsaw: Polish Scientific Publishers.
Plekhanov, G. V. (1976). *Selected philosophical works II.* Moscow: Progress Publishers.
Sommerfeld, A. (1917). Zum Andenken an Marian von Smoluchowski (In Memory of Marian von Smoluchowsk). *Physikalische Zeitschrift*, *22*(15), 533–539.
von Plato, J. (1994). *Creating modern probability, its mathematics, physics and philosophy in historical perspective*. Cambridge University Press.
von Smoluchowski, M. (1913). Gültigkeitsgrenzen des zweiten Hauptsatzes der Wärmetheorie [Limits to the validity of the second law of thermodynamics]. *Physikalische Zeitschrift*, *15*, 261.
von Smoluchowski, M. (1914). Gültigkeitsgrenzen des zweiten hauptsatzes der wärmetheorie (Limits to the validity of the Second law of thermodynamics). In M. Planck (Ed.), *Vorträge Über Kinetische Theorie der Materie und Electrizität* (pp. 87–121). Leipzig: Teubner.
von Smoluchowski, M. (1916). Drei Vorträge über Diffusion, Brownsche Molekularbewegung und Koagulation von Kolloidteilchen [Three lectures on diffusion, brownian molecular motion and the coagulation of colloidal particles]. *Physikalische Zeitschrift*, *17*(23), 557–571.
von Smoluchowski, M. (1918). Über den Begriff des Zufalls und den Ursprung der Wahrscheinlichkeitsgesetze in der Physik [On the concept of chance and the origin of the laws of probability in physics). *Die Naturwissenschaften [The science of nature]*, *6*(17), 253–263.

Chapter 6
Mechanical Materialism and Modern Physics (Section 1)

Boris Hessen

The debate between the Dialecticians and Mechanists will soon celebrate its five-year anniversary. *Dialectical Materialism and the Deborin School*, a book by Comrade Stepanov,[1] is a truly celebrated achievement in its form if not its contents. One cannot continue polemics in the spirit of this book, and not only because it has exhausted the entirety of printable abuse. The truth of a point of view is decided by the methodological analysis of a scientific problem and the historical achievements of science, and not by the "strength" of words used to support it. Unfortunately, Comrade Stepanov's final essay contains even less concrete material than the previous ones and is not, as we shall try to demonstrate, on a par with the level of modern science, whose ardent defender Comrade Stepanov is. We see the main drawback of the book in its careful avoidance of all burning issues of modern natural science while making a hearty defence (who from?) for example, of the law of conservation of energy that has for a long time undisputedly been part of the arsenal of natural science.

[1] TN: For more on Ivan Ivanovich Skvortsov-Stepanov, to give him his full name, see the Appendix. The book was published by Gosizdat, Moscow-Leningrad, 1928. Stepanov died of typhoid in October of the same year. This article by Hessen was referred to as a first part but a second part never appeared, presumably because of Stepanov's death.

TN: Translated from *Pod Znamenem Marksizma* (*Under the Banner of Marxism*), 1928, Nos 7–8. The article was in three sections. This chapter contains the first section, the next two chapters contain the second and third sections.

B. Hessen (✉)
Abingdon, UK
e-mail: chris.talbot@protonmail.com

C. Talbot and O. Pattison (eds.), *Boris Hessen: Physics and Philosophy in the Soviet Union, 1927–1931*, History of Physics,
https://doi.org/10.1007/978-3-030-70045-4_6

At the same time Comrade Stepanov substitutes recent problems by abuse and angry accusations. However, swearing is a poor means for solving problems and an indirect proof of the realisation of one's unjust position: "Jupiter, you are angry, therefore, you are wrong".[2]

We shall not follow Comrade Stepanov's style of polemics but shall try to find our point of view by examining concrete material and to oppose it with the point of view of the Mechanists.

Let us turn for this purpose to the analysis of classical and modern natural science but limit our task to physical problems.

1. The mechanical and 'mechanistic' worldview.

According to Comrade Stepanov dialectical materialism in natural science is concretised as mechanistic materialism. Dialecticians purposefully mix pre-chemical mechanical materialism with the "mechanistic" materialism of the end of the nineteenth—beginning of the twentieth century, which is essentially dialectical materialism.[3]

What is the real situation? Is modern natural science "mechanistic"? What is the difference between the mechanical worldview of the eighteenth century and the mechanistic (according to Comrade Stepanov's terminology) natural science of the end of the nineteenth century? In order to decide whether there is a real difference between the mechanical and mechanistic view on nature and what is the methodological essence of the mechanical worldview we shall look at the works by those who created natural science in the nineteenth century (therefore, not pre-chemical). What was the essence of the mechanical worldview according to Helmholtz, Maxwell, Boltzmann and du Bois-Reymond? How did they define the task and goals of the mechanical worldview? Unfortunately, Comrade Stepanov who strives to defend modern natural science, albeit in his essays, fails to demonstrate sufficient knowledge of the true views of the founders of the mechanical approach in natural science. We shall try to examine the views of these founders of nineteenth century science although it might be a little more difficult than to read an article in an encyclopaedia. We hope Comrade Stepanov will agree with us that such articles, although they save a lot of time and do not require deep knowledge to find, are insufficient for a scientific solution of a problem. Therefore, let us look at the original works by the above-named natural scientists.

The mechanical worldview arose as a reaction against the scholastic physics of hidden qualities.

Scholastic physics called hidden those qualities that cannot be perceived and studied but are the causes for the observed phenomena. A magnet attracts iron because it has the magnetic gravity force (this force being the hidden quality *qualitas occulta*).

---

[2]TN: Latin proverb, well known from Dostoevsky, *The Brothers Karamazov*.

[3]BH: See "Dialectical Understanding of Nature – Mechanistic Understanding" in the collection *Dialectical Materialism and the Deborin School*. TN: i.e. the book referred to in the opening paragraph.

Molière's doctor replies to the question of why opium helps one to sleep, "because it contains sleep making properties whose nature is to induce drowsiness in the senses".[4]

This is essentially the "scientific" explanation of phenomena by scholastics. Clearly this methodology could not be used as a scientific tool and was forcefully opposed by Descartes.

> For I openly acknowledge that I know of no kind of material substance other than that which can be divided, shaped and moved in every possible way, and which Geometers call quantity and take as the object of their demonstrations. And [I also acknowledge] that there is absolutely nothing to investigate about this substance except those divisions, shapes, and movements; and that nothing concerning these can be accepted as true unless it is deduced from common notions, whose truth we cannot doubt, with such certainty that it must be considered as a Mathematical demonstration. And because all-natural phenomena can be thus explained, as will appear in what follows; I think that no other principles of Physics should be accepted or even desired.[5]

These words describe the entire programme of the mechanical worldview: all natural phenomena must be explained and reduced to the motion and position of elementary particles.

E. Whittaker in his famous work *A History of the Theories of Aether and Electricity* says, "The grandeur of Descartes's plan, and the boldness of its execution, stimulated scientific thought to a degree before unparalleled; and it was largely from its ruins that later philosophers constructed those more valid theories which have endured to our own time."[6]

The mechanical worldview formulated by Descartes became a lodestar in the natural science of the seventheenth and eighteenth centuries, and a philosophical credo of natural scientists.

Huygens said, "... true Philosophy, in which one conceives the causes of all natural effects in terms of mechanical motions. This, in my opinion, we must necessarily do, or else renounce all hopes of ever comprehending anything in Physics."[7]

The primary task of mechanical physics was to destroy the metaphysical view of nature as a set of manifestations of specific hidden qualities. All natural phenomena should be explained using a single principle, i.e. the motion and position of elementary particles.

Descartes ruthlessly battled against the scholastic physics of hidden qualities and did not stop before reducing mass to extension. Every other understanding of mass was a hidden quality for him.

---

[4]TN: Molière, *Le Malade Imaginaire*. The original is in Latin: "Quia est in eo/ virtus dormitiva/ cujus est natura/ sensus assoupire.".

[5]TN: Hessen references a Russian translation. The text used here is taken from the English translation in Descartes (1983), Part II, paragraph 64, p. 77.

[6]BH: Whittaker (1910, p. 3).

[7]BH: Huygens, Traité de la Lumière, Paris, 1910, p 3. TN: English translation used here in Huygens (1969, p. 3).

Cartesians fiercely opposed Newton precisely because he additionally introduced forces acting between elementary particles into the formulation of the mechanical worldview which explained all phenomena on the basis of the motion and position of elementary particles.

Descartes considered a force acting according to a certain law (gravitation) but not reduced to the motion and position of elementary particles to be a hidden quality. In the second part of *Principles* he attempted to explain cohesive and gravitational forces by the motion and position of particles.

Although the gravitational forces could not and still cannot be explained mechanically, a purely phenomenological concept of force introduced by Newton proved to be so fruitful that it has firmly joined the toolbox of exact natural science. The formulation of the mechanical worldview was complemented by the concept of force, and the explanation of all natural phenomena on the basis of the motion and position of elementary particles and forces acting between them became the task of physics.

Newton formulated the tasks of the mechanical worldview in the eighteenth century while Laplace extended the law of gravitation to the micro-world by the mechanical theory of capillary phenomena. The greatest achievements of physics in the nineteenth century are linked to the triumph of the mechanical worldview (the Fresnel mechanical wave theory of light) and found their final expression in kinetic gas theory.

At the same time the weak aspects of this formulation became more noticeable.

We saw earlier how Descartes defined the task and essence of the mechanical worldview. This definition is wholly unconnected with the concrete state of mechanics at the time. The problem is not in the laws of motion of elementary particles or whether their motion can be described by Newton's equations or Lagrange's equations. The mechanical worldview of Descartes is a general methodological premise whereby all properties of matter have to be explained *only* by the motion and position of elementary particles of matter. The explanation of the world for him is a purely kinematic problem.

The forces acting between the particles are already a compromise because they are yet unexplainable by the same principle.

Consequently, the development of chemistry and biology could not have an impact on the fundamental formulation of the mechanical worldview. It was certainly possible to introduce electrical forces, chemical bonding forces etc. along with the gravitational ones acting between the particles but until these new properties of elementary particles (molecules, atoms or electrons) could be explained as motion and position of elementary particles the task of the mechanical worldview remained unfulfilled.

But what are these elementary particles and which properties are really elementary and could not be reduced further?

Descartes replied to this question by declaring extension, i.e. volume, as the only property of elementary particles.

> "The nature of body does not consist of hardness which affects our senses. Nor is it in its weight, heat or other similar properties; for if we deconstruct a body we can always imagine that it does not have any of these properties, while at the same time we clearly and definitely recognise that it has everything that makes it a body, as long as it possesses a length, a breadth

and a depth. Therefore, for its being it has no need in them, and its nature is in its substance which has extension."—is Descartes's view on the nature of elementary particles.[8]

Comrade Stepanov will probably retort that these views belong to pre-chemical mechanical materialism and that the "mechanistic" materialism of the second half of the nineteenth century is completely different.

Let us have a look. It is hard to deny that du Bois-Reymond, one of the brilliant representatives of mechanical worldview in natural science does not represent "pre-chemical materialism".

Here are his views on the essence of mechanical worldview:

> For us there is no other method of cognition apart from the mechanical one (mechanische), therefore, the physical and mathematical form of thinking is the only scientific one. . . . theoretical natural science will be satisfied only when it reduces (zurückführt) all phenomena to motions of elementary particles which occur under the same laws as in the ruder sensual matter (gröberen sinnfälligen Materie)[9]

We see that there is no fundamental difference between the formulations by du Bois-Reymond and Descartes. Nor could there be as Descartes provided a general methodological premise. The problem of the mechanical worldview is *fundamentally* the same at the end of the nineteenth and in the middle of the seventeenth century.

The following two main features in the du Bois-Reymond definition are important:

1. He does not mention "chemical forces and processes" at all and only speaks about motion of elementary particles.

2. He states that laws of the micro and macro-world are identical. This is one of the fundamental premises of mechanical natural science and a governing principle of exact physics in the nineteenth century.

Of course it is not accidental that du Bois-Reymond does not introduce chemical forces into his formulation of the mechanical worldview. He understands that according to the mechanical view elementary particles should not have any other properties

---

[8]TN: This is a translation from Hessen's Russian text. It appears to be a loose translation of Descartes (1983, p. 40): "That the nature of body does not consist in weight, hardness, color, or other similar properties; but in extension alone. By so doing, we shall perceive that the nature of matter, or of body considered in general, does not consist in the fact that it is hard, heavy, colored, or affects the senses in any other way; but only in the fact that it is a thing possessing extension in length, breadth, and depth," and Descartes (1983, p. 41): "Therefore, the nature of body does not consist in hardness. In the same way, it can be shown that weight, color, and all the other properties of this kind which are experienced in material substance, can be taken away; leaving that substance intact. From this it follows that the nature of matter does not depend on any such properties but consists solely in the fact that it is a substance which has extension."

[9]BH: Reden, I., p. 464. We would like to note that his mechanical worldview did not prevent du Bois-Reymond from being agnostic and sometimes leaning towards idealism. TN: A translation from Hessen's Russian translation from the German is given. The German original is from two sources: "Es giebt für uns kein anderes Erkennen, als das mechanische, ein wie kümmerliches Surrogat für wahres Erkennen es auch sei, und demgemäss nur Eine wahrhaft wissenschaftliche Denkform die physikalisch-mathematische." du Bois-Reymond (1886, p. 232). and ". . . die theoretische Naturwissenschaft ruht nicht eher, als bis sie die Erscheinungswelt auf Bewegungen letzter Elemente zurückführte, welche nach denselben Gesetzen vor sich gehen, wie die der gröberen, sinnfälligen Materie." du Bois-Reymond (1886, p. 434).

apart from the ability to move in space. This is why du Bois-Reymond calls them *without properties* when defining the properties of these elementary particles.

> The mechanical worldview which knows only oscillations of primary matter without properties instead of sound and light and is obtained by means of objective observation sees the world as silent and dark, i.e. *property-less*.[10]

Isn't it strange that du Bois-Reymond speaks in "Deborin" language about particles without properties, but we leave the solution of this problem to the Timiryazev Research Institute.

There is as we see, nothing fundamentally new in du Bois-Reymond's formulations compared to the pre-chemical period. The main principle, i.e. the explanation of all phenomena on the basis of mechanical motion and the position of elementary particles remains the scope of mechanics.

Let us look at the views of the leading physicists of the nineteenth century.

Maxwell, the founder of heat theory, kinetic gas theory and electromagnetic field theory begins his "Essay on modern molecular physics, in particular, molecular gas theory"[11] with the following definition of the tasks of molecular physics:

> We begin by assuming that bodies are made up of parts, each of which is capable of motion, and that these parts act on each other in a manner consistent with the principle of the conservation of energy. . .
> We may also assume that these small parts are in motion. This is the most general assumption we can make, for it includes, as a particular case, the theory that the small parts are at rest. . . We make no assumption with respect to the nature of the small parts . . . We do not even assume them to have extension and figure. Each of them must be measured by its mass, and any two of them must, like visible bodies, have the power of acting on one another when they come near enough to do so. *The properties of the body, or medium, are determined by the configuration and motion of its small parts.*[12]

H. von Helmholtz gives an identical formulation in his study on conservation of force:

> We must not attribute qualitative differences to matter itself because when we speak of different types of matter, we assume the difference being only in actions, i.e. their forces. The matter itself, therefore, cannot undergo any other change but a spatial one, i.e. motion.[13]

---

[10]BH: Uber die Grenzen d. Naturerkenntniss, 1916, p. 22. TN: Again a translation from Hessen's Russian translation from the German is given. It appears to be a loose translation of the passage from "The Limits of our Knowledge of Nature" translated into English in du Bois-Reymond (1874) as: "And voiceless and dark in itself, i.e., property-less, as the universe is on subjective decomposition of the phenomena of sense, so is it also from the mechanical stand-point, gained by objective contemplation. Here, in place of sound and light, we have only the vibrations of a primitive, undifferentiated matter, which here has become ponderable, and there imponderable."

[11]BH: *Speeches and works by James Clark Maxwell*, 1901, Moscow, publication and translation by Marakuyev, p. 58. TN: The original English "Outline of Modern Molecular Science, and in particular of the Molecular Theory of Gases", Niven (1965, p. 451) is given here.

[12]TN: Italics added by Hessen.

[13]TN: German original: "Qualitative Unterschiede dürfen wir der Materie an sich nicht zuschreiben, denn wenn wir von verschiedenartigen Materien sprechen, so setzen wir ihre Verschiedenheit immer nur in die Verschiedenheit ihrer Wirkungen d. h. in ihre Kräfte. Die Materie an sich kann deshalb

One of the main difficulties of the mechanical worldview—the problem of the properties of an elementary particle—is most clearly revealed in Maxwell's definition. Indeed, no properties should be assigned to this elementary particle as all properties of these particles should be explained on the basis of their motion and position. Descartes identified a particle's mass with its extension, i.e. used kinematics as the basis for the explanation of a body's properties. His particles are characterised only by volume. Here Maxwell could not follow Descartes. His particles should be measured by their mass, but he assigns them *neither extension nor form.*

H. von Helmholtz expressed these difficulties for elementary particles even clearer in his speech about Gustav Magnus:

> In reference to atoms in molecular physics Sir W. Thomson says with much weight that their assumption can explain no property of the body which has not previously been attributed to the atoms. Whilst assenting to this opinion, I would in no way express myself against the existence of atoms, but only against the endeavour to deduce the principles of theoretical physics from purely hypothetical assumptions as to the atomic structure of bodies. . . . In our immediate experience we find bodies variously formed and constituted, only with such can we make our observations and experiments Their actions are made up of the actions which each of their parts contributes to the sum of the whole; and hence, if we wish to know the simplest and most general law of the action of masses and substances found in nature upon one another . . . we must go back to the laws of action of the smallest particles, or, as mathematicians designate it, the elementary volume. But these are not, like the atoms, disparate and heterogeneous, but continuous and homogeneous.[14]

Helmholtz clearly sees the difficulties of atomic theory. Atoms for him are identical smallest particles whose motion and position should explain all properties of a body. He rebels against any hypothesis regarding the composition of elementary particles (atoms). For him they are only particles of volume. He says nothing at all about reducing all phenomena to physical and chemical phenomena because by accepting atomic physical and chemical properties one assigns them the properties that require an explanation.

The mechanical measure of explanation for him, as well as for Maxwell, W. Thomson and du Bois-Reymond, is the only one.

He says in his speech "On the purpose and achievements of natural science",

> . . . the ultimate aim of physical science must be to determine the movements which are the real causes of all other phenomena and discover the motive powers on which they depend; in other words, to merge itself into mechanics.[15]

Helmholtz made this speech in 1869.

Comrade Stepanov might call this time also pre-chemical.

Let us then look at statements made by William Thomson in 1898. W. Thomson was one of the greatest physicists of the nineteenth century and the founder of vortex

auch keine andere Veränderung eingehen, als eine räumliche, d.h. Bewegung." Helmholtz (1847, p. 3) (Introduction).

[14]TN: English translation from "Gustav Magnus, In Memoriam" in Helmholtz (1884, pp. 17–18).

[15]BH: Vortrage, p. 375, Russian translation in the collection "Philosophy of Science", Physics, pt.1, p. 51. TN: English translation from "The Aim and Progress of Physical Science" in Helmholtz (1995, p. 211).

atomic theory. He dedicated his life to the construction of a mechanical theory of matter. After the failure of the grandest of his ideas, the vortex atomic theory, he wrote,

> I am afraid that it is not possible to explain all the properties of matter by the vortex-atom theory alone, that is to say, merely by motion[16] of an incompressible fluid ; and I have not found it helpful in respect to crystalline configurations, or electrical, chemical, or gravitational forces . . . We may expect the time will come when we shall understand the nature of an atom. With great regret I abandon the idea that a mere configuration of motion suffices.[17]

Therefore, at the turn of the twentieth century the explanation of all body properties (chemical, electrical or gravitational) by mechanical motion and the configuration of discrete particles or fluids was the governing idea for natural scientists.

Let us draw some conclusions on the basis of the above quotations.

The distinction between mechanical and mechanistic worldviews made by Comrade Stepanov does not stand up to criticism. There was and still is one mechanical worldview that, as we have seen, ruled in nineteenth century natural science.[18]

When Comrade Stepanov accuses Comrade Sten in "stubbornly sticking to the mechanical view, i.e. from chemistry and physics always sliding into mechanics",[19] Comrade Sten can be justified in his understanding of the essence of the mechanical worldview by citing identical views of Helmholtz, Maxwell, W. Thomson and du Bois-Reymond.

There is no difference between the mechanical and "mechanistic" worldview but there is a difference between the mechanical worldview and dialectical materialism. Therefore, every reproach Engels addressed to mechanical materialism is valid for the entire nineteenth century natural science.

We have seen that the application of the body of knowledge of mechanics to chemical and biological phenomena, i.e. what Engels reproached the old materialists for, is common to all natural scientists of the second half of the nineteenth century, who took the conception of mechanical materialism.

And it is certainly no coincidence. Du Bois-Reymond, Helmholtz, Maxwell and W. Thomson leave information on physical and chemical phenomena out of their reducibility formulas not because they are unfamiliar with chemistry but because, in order to be consistent, no problem should be formulated unless *all phenomena are reduced to mechanics.*

If the above reduction is recognised as the governing principle of scientific research, then one should not dwell on physical and chemical phenomena. However, if physical and chemical phenomena are recognised as elementary and everything is reduced to them while they themselves are not reduced to mechanics, then a certain part of phenomena should be recognised as irreducible, and therefore, the reduction

---

16BH: i.e. mechanical

17BH: Letter to Silas W. Holman, 1898, in Thompson (2005, p. 1047, n 1).

18BH: These quotations cover W. Thomson's statements between 1847 (the date of Helmholtz's essay on conservation of forces) and 1898.

19BH: Stepanov, *Dialectical Materialism and the Deborin School*, p. 97.

principle should be fundamentally abandoned. It is this inconsistency that Comrade Stepanov uses to support his difference between the mechanical and "mechanistical".

This is why all criticism of mechanical natural science by Dialecticians is valid.

Certainly, we do not criticise mechanical materialism for being materialistic but for its general methodological task of reducing all phenomena to mechanics and for thereby closing the door to the possibility of studying natural processes which do not fit into the mechanical framework. According to Lenin, it was impossible to "develop the theory of materialism" following this approach. The development of exact natural and biological science required the development of the theory of materialism. Science faced new problems unsolvable within the framework of mechanical materialism. This brought about the crash of the mechanical worldview noticeable already in the 1880s. Earlier we saw W. Thomson's abandonment of the construction of a mechanical theory of matter (vortex-atom). Attempts to construct mechanical theories of electromagnetic fields also failed.

Hertz's mechanics is the last and the most grandiose attempt to reduce all natural phenomena to mechanical movement. M. Planck said, ". . . the search of the mechanical conception for a uniform world picture has been brought to a somewhat ideal completion. Hertz's mechanics is not physics of today but physics of tomorrow, or as it were, a sort of confession of faith for physics."[20]

This "confession of faith" means that H. Hertz believed it possible to "fully explain the mechanical point of view by assuming movements of simple homogeneous material points, i.e. the only true building blocks of the universe."[21]

As is known Hertz's attempt also failed and had no further influence on the development of natural science.

All right, Comrade Stepanov might argue, maybe all these failures can be explained because the scientists disregard the achievements of chemistry and biology and are reluctant to adopt the mechanistic standpoint, while rising above the mechanical picture. In this case, however, he should explain why *all* natural scientists, from Helmholtz and du Bois-Reymond to Hertz speak about the reduction of all phenomena to mechanical movements of a simple homogeneous point. Not only were they all armed with the knowledge of the second half of the nineteenth century but themselves were the founders of this scientific development.

---

[20]BH: M. Planck, *Physical essays*, p. 39. TN: It is not clear which source Hessen is using. A translation from the Russian is given here. The original is in Planck (1960, p. 31): ". . . the search of the mechanical conception for a uniform world picture has been brought to a somewhat ideal completion. Hertz's mechanics does not really represent physics as it is, it is physics as it might be, a sort of confession of faith for physics."

[21]BH: Same, p. 39. Readers will notice that both M. Planck and Hertz speak "Deborin's language" about simple homogeneous material points. TN: Again a translation from the Russian original is given. An English version is in Planck (1960, pp. 31–32): ". . . all Nature, from the mechanical point of view, can be completely explained by assuming movements of simple, similar particles, which build up the whole of the physical universe."

Indeed, there was no alternative: either consistent mechanism, i.e. everything should be reduced to the motions of a homogeneous material point, or inconsistent mechanism, which according to W. Thomson, endows atoms with material properties that have to be explained (chemical or electrical).

The failure of mechanical theories brought about a renaissance of idealistic trends in physics. Many philosophers or philosophising natural scientists began looking for a solution to the crisis by "overcoming natural scientific materialism" instead of further developing the theory of materialism.

Comrade Stepanov would like to picture our criticism of mechanical materialism as identifying with the trend which sees the drawbacks of mechanical materialism in its materialistic character. We pay due tribute to the great achievements of the mechanical worldview in the nineteenth century. We are fully aware of the fact that the natural scientists of the time identified the mechanical with the materialistic. However, this does not mean that in our time one has to share mechanical materialism's standpoint in order to be a materialist in natural science. It is possible to be one if one rises to the standpoint of dialectical materialism. We shall repeat after R. Millikan, "We can still look with a sense of wonder and [respect] and reverence upon the fundamental elements of the [physical] world as they have been partially revealed to us in this century. The childish mechanical conceptions of the nineteenth century are now grotesquely inadequate."[22]

Indeed, the achievements of physics, chemistry and biology forced natural scientists to review the mechanical methodology but not in the way described by Comrade Stepanov.

For him the application of the energy conservation law to all natural phenomena including the psychological ones, is the main victory of the "mechanistic" conception. He even identifies energy conservation with the "mechanistic worldview".[23]

Undoubtedly, the law of conservation of energy is a mighty weapon for banishing all mystical forces from natural science. However, it is quite wrong to identify it with the mechanistic conception. This is Comrade Stepanov's reasoning: Dialecticians deny the "mechanistic worldview", that is essentially none other than the law of energy conservation. Therefore, they deny the energy conservation law thus opening little doors for different mystical forces as far as a life force,[24] which have been banished from natural science forever.

To start with, a historical note: the energy conservation law was clearly worded by Robert Mayer in 1845 in his study "The organic movement in connection to metabolism"[25] and in 1847 Helmholtz generalised its application to all phenomena.

---

[22]BH: An essay by R. Millikan in *Scientia*, 1926. TN: Original English from Millikan (1927).

[23]BH: "The energy conservation (and transformation) law that I defend under the name of the mechanistic understanding of nature . . . ." (Stepanov, "Dialectical materialism and the Deborin school", p. 96).

[24]TN: i.e. vitalism.

[25]TN: i.e. Robert Mayer's 1845 article: "Die organische Bewegung in ihrem Zusammenhange mit dem Stoffwechsel. Ein Beitrag zur Naturkunde.".

The mechanical theory of heat did not exist at that time. The first works by Kronig and Clausius date to 1856. No attempts were yet made to support electromagnetic phenomena by a mechanical theory.

The historical law of energy conservation had been scientifically founded before the wide application of the mechanical approach in molecular physics. It is simply wrong to share Comrade Stepanov's belief that the law of the conservation of energy is identical with the mechanical worldview. This contradicts the historical development of natural science. As we have seen earlier, the mechanical worldview had been scientifically formulated by Laplace early as the end of the eighteenth century when the law of conservation of energy was yet unknown.

Undoubtedly, the general premise of the mechanical approach played an important part in the history of the energy conservation law. And not without reason Descartes formulated this law simultaneously with his mechanical picture of the world.[26] But this does not justify putting an equal sign between them, as does Comrade Stepanov.

The mechanical approach has developed and enriched the scientific formulation of the energy conservation law, and this law remains the main governing principle of natural research although the mechanical (or mechanistic) approach is not at all shared by modern physicists.

The law of energy conservation (and transformation) played a hugely revolutionary role in natural science. It united all types of energy that were deeply divided before. It was a mighty weapon in the destruction of the metaphysical view of nature and in the establishment of the connection between all organic and non-organic natural phenomena.

However, the energy conservation law is not sufficient in spite of its significance. Comrade Stepanov forgets that the foundation of natural science includes the law of the dissipation of energy as well as of its conservation.

While the law of energy conservation emphasizes the unity of all phenomena, the law of energy dissipation establishes their specificity and introduces a fundamentally new approach to science. We think it is not accidental that Comrade Stepanov dismisses the latter law. While he justly confirms the great significance of the law of conservation of energy for the dialectical approach to nature, Comrade Stepanov forgets that the dialectical approach is in establishing both *the unity and the particularity*. This is why he ignores the law of energy dissipation which establishes specific differences between purely mechanical systems and those that, although governed by mechanical laws in their components, have specific differences, i.e. specific laws.

We shall attempt to show that the mechanical worldview is insufficient even within the *framework of physics*; that the reduction to mechanics is impossible even in the mechanical theory of heat, i.e. kinetic gas theory. But first, a few words on specificity. According to Comrade Stepanov, "... to say'specificity', to say that it is irreducible to physical and chemical processes and to stop there is tantamount to agreeing that some phenomena in life are unknowable."[27] However, we would be grateful to Comrade

---

[26] BH: Descartes formulated the law of the conservation of energy as a general philosophical premise, as a law of the conservation of motion.

[27] BH: Stepanov, p. 12.

Stepanov for showing us when did Dialecticians propose to stop at specificity. We affirm that no one has ever said or written this. We have always emphasized the problem of particularity together with the universality of phenomena because science inevitably faced the former as soon as unity of phenomena was established, e.g. by the law of energy conservation. It is not enough to say that light, heat and electricity are different types of energy; it is essential to establish their specific differences as well as their unity. It is not enough to say that heat is a mechanical motion of molecules or that the volume of gas consisting of a vast number of molecules is a mechanical system; it is essential to point out the specific differences between this system and a simple aggregate of separate molecules moving under the laws of mechanics and not limited thereto. And these specific differences do exist, they are irreducible to the laws of mechanics and it was about them Engels wrote,

> . . . that heat is a molecular motion . . . But if I have nothing more to say of heat than that it is a certain displacement of molecules, I should best be silent.[28]

Comrade Stepanov believes that if one accepts specificity and unity, then one is a reactionary scientist.[29] However, when criticising the bourgeois economists, Marx emphasized that "It is necessary to distinguish those definitions which apply to production in general, in order not to overlook the essential differences existing despite the unity . . . on failure to perceive this fact depends the entire wisdom of modern economists . . ."[30]

We prefer to study Marxist methodology according to Marx and not to the entries in the Soviet Encyclopaedia even if written by outstanding natural scientists.

References

Descartes, R. (1983). *Principles of philosophy* (V. R. Miller & R. P. Miller, Trans.). Dordrecht: Reidel.

du Bois-Reymond, E. (1874). The limits of our knowledge of nature. *Popular Science Monthly, 5*. (J. Fitzgerald, A. M. Trans.).

du Bois-Reymond, E. (1886). *Reden von du Bois-Reymond*. Leipzig: Von Veit and Co.

Engels, F. (1988). *Marx Engels collected works* (Vol. 25). Moscow: Progress Publishers.

Helmholtz, H. (1847). *Über die Erhaltung der Kraft, eine physikalische Abhandlung, vorgetragen in der Sitzung der physikalischen Gesellschaft zu Berlin am 23sten Juli 1847*. Berlin: G. Reimer.

Helmholtz, H. (1884). *Popular lectures on scientific subjects* (E. Atkinson, Trans.). London: Longmans, Green and Co.

Helmholtz, H. (1995). In D. Cahan (Ed.), *Popular and philosophical essays*. Chicago: University of Chicago Press.

Huygens, C. (1969). *Treatise on light*. New York: Dover.

Millikan, R. (1927). Conceptions in physics changed in our generation. *Scientia, 41*(180), 255–264.

---

[28]TN: See *Dialectics of Nature*, Engels (1988, p. 531).

[29]BH: Stepanov, "Dialectical materialism and the Deborin school", p. 103.

[30]BH: K. Marx, A contribution to the critique of political economy, Moskovski Rabochi, 1923, p. 11. TN: https://www.marxists.org/archive/marx/works/1859/critique-pol-economy/appx1.htm. The English translation is used here.

Niven, W. D. (Ed.). (1965). *The scientific papers of James Clerk Maxwell* (Vol. II). New York: Dover.
Planck, M. (1960). *A survey of physical theory*. New York: Dover.
Thompson, S. (2005). *Life of Lord Kelvin* (Vol 2.). American Mathematical Society – Chelsea, Providence, Rhode Island. Reprint of Macmillan (1910).
Whittaker, E. (1910). *A history of the theories of Aether and electricity*. London: Longmans, Green and Co.

Chapter 7
Mechanical Materialism and Modern Physics (Section 2)

Boris Hessen

II. What is the specificity of physical phenomena and their irreducibility to mechanics?

Let us closely examine the law of energy dissipation and its kinetic interpretation in order to clarify the problem of specificity and irreducibility of phenomena. This would allow us to clarify our essential formulations on concrete material. We shall deal with a purely physical law but the fundamental content of our formulations becomes sufficiently clear in physics as well [as in other areas].

A volume of a gas consists of a vast number of constantly moving molecules. Therefore, gas can be considered as a system of molecules moving according to the laws of mechanics. Let us look at a law applicable to gases as a whole, e.g. Boyle's law, whereby the gas pressure on a vessel's walls is inversely proportional to its volume. Clearly this law makes sense only in relation to the gas volume as a whole because the concept of pressure changing with changes in volume makes no sense with regards to a single isolated molecule or, in any case, has a completely different meaning since we assume that the volume of a *single* molecule does not depend on the pressure of the entire gas when we make an assumption of the molecular composition of the gas.

The motion of a *single* molecule is a purely mechanical process and is fully described by mechanical equations.

What does it mean that the gas pressure and Boyle's law may be reduced to mechanical molecular motion?

TN: Translated from *Pod Znamenem Marksizma* (*Under the Banner of Marxism*), 1928, Nos 7–8.

B. Hessen (✉)
Abingdon, UK
e-mail: chris.talbot@protonmail.com

C. Talbot and O. Pattison (eds.), *Boris Hessen: Physics and Philosophy in the Soviet Union, 1927–1931*, History of Physics,
https://doi.org/10.1007/978-3-030-70045-4_7

It means that by establishing purely mechanical laws of motion of single molecules the laws for gases as a whole can be deduced from these premises, and *only from them*. The laws for gases as a whole (e.g. Boyle's law) are a simple arithmetic sum of mechanical laws governing the movement of single molecules. No specific laws for gases as a whole compared to the laws for the gas components have been observed.

If this is the case we can say that the gas pressure on a vessel's walls is reduced to the mechanical impact of molecules constantly bombarding these walls, and Boyle's law is reduced to mechanical laws of motion of molecules.

Is this reduction of thermodynamic laws to mechanics possible in kinetic gas theory? Comrade Stepanov states that such a reduction fundamentally takes place. If all biological phenomena can be reduced to physical or chemical ones then according to Comrade Stepanov, even more so purely physical phenomena can be reduced to mechanical ones. Anyone who argues otherwise is a reactionary scientist and a "physical vitalist".

Let us look at what really takes place.

I throw a stone from a high place. A stone raised to a certain height has a certain reserve of (potential) energy. When falling to the ground the stone produces a certain amount of heat (heat energy) and a sound, a disturbance of the air, i.e. a certain amount of mechanical energy, a known deformation of the soil, etc.

According to the law of energy conservation the amount of all the energy produced by a falling stone, i.e. the energy into which the potential energy stored in a stone lifted to a certain height is transformed, should be equal to the potential energy of the stone.

However, we see here that apart from the equal amount of energy we deal with a process proceeding *in a certain direction*. *At first* there is potential energy that is *later* transformed into a number of other energies. The process, as it were, proceeded from mechanical potential energy to heat and other energies. The stone was first raised to a certain height and then, having fallen, produced other types of energy. Is this direction of the process obligatory? The stone produced a certain amount of heat energy after its fall that warmed the soil and the surrounding air. Can it not raise itself back to the previous height *by the use of this heat*[1] and accordingly cooling the air and earth?

In other words, can the transformation of energy proceed in reverse by itself?

When we say "by itself", we mean without an intervention by extraneous forces as happened when the transformation of potential energy took place spontaneously without such an intervention.

The law of the conservation of energy does not give a reply to this question but experiments give a negative one. Never and nowhere in nature did we observe such a transformation. The process proceeds in a strictly determined direction and cannot be reversed. We can divide all natural processes into *reversible and irreversible* ones from the point of view of the *direction*. A pendulum's swing is the simplest and most typical example of a reversible process.

[1] BH: We disregard other types of energy due to their insignificant amount compared to heat energy.

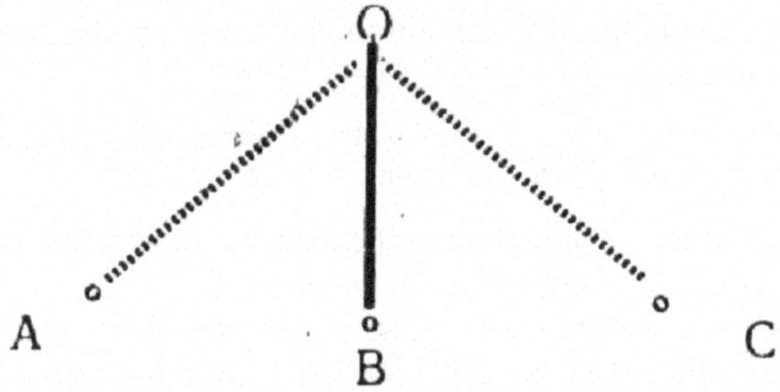

If we swing a pendulum out of equilibrium OB into the position A and then leave it alone, it will swing between A and C all the time. Its potential energy will transform into kinetic and vice versa. If we disregard the pendulum's air resistance and the friction in the supporting string, then the swings related to the transformation of energy from potential to kinetic will continue for an indefinite time, first in one direction and then in the other.

The process is quite reversible.

Indeed, any mechanical process we choose is always reversible.

However, *thermal processes* are quite different. Let us drop a warm metal ball into a glass of cold water. Heat[2] will transfer from the ball to water. The water will warm up and the ball will cool down. The amount of heat energy lost by the ball will be equal to that acquired by the water.

The heat transfer from the ball to the water will end here. Heat is transferred from a hotter body to a cooler one. At the end of the process the same temperature (i.e. the thermal equilibrium) of the ball and water is established. However, the process does not proceed in reverse.

The pendulum passed through the equilibrium position OB and went further. From A to B it *went down*, and having passed the equilibrium point B, it started *going up*. The heat process did not go beyond the equilibrium point. Its flow is strictly one-way. As in the example with a stone, the process can proceed only in one direction: always and everywhere we note the transfer of heat from a hotter to a cooler body. We have never observed a reverse transfer. *Thermal processes are irreversible*.

It seems, reversible and irreversible processes are divided by an impassable gap.

The direction of reversible processes is not determined. They can proceed in any direction. Conversely, the direction of irreversible processes is strictly determined.

The second law of thermodynamics establishes this difference between reversible and irreversible processes. This law compliments the first law of thermodynamics (the energy conservation law) by pointing to the direction of transformation. The content of the law can be worded in different ways. Below are the two most popular formulations. "Heat cannot spontaneously transfer from a cooler to a hotter body."[3] This wording by Clausius clearly points at a certain direction of heat processes. We have essentially looked at this formulation in the example of a ball heating water.

[2]BH: There is no need for us here to go into the essence of heat.

[3]TN: From Clausius, "On a Modified Form of the Second Fundamental Theorem in the Mechanical Theory of Heat" (1856). See also Clausius (1879), p. 212.

Another formulation belongs to W. Thomson and points out the difference between mechanical and thermal processes.

> It is impossible to construct an engine which will work in a complete cycle, and produce no effect except the raising of a weight and cooling of a heat reservoir.[4]

We have looked at this formulation in the example of a falling stone. No device can lift a fallen stone by cooling its environment.

The impossibility of building an engine under Thomson's second law, the so-called perpetuum mobile of the second type, does not follow from the principle of energy conservation. Indeed, if it were possible to raise a stone due to the cooling of the medium, we would have achieved the transformation of heat energy into the (potential) energy of the elevated stone. The amount of energy would have been equal, and the law of energy conservation would have been observed.

A statement about the impossibility of such a direction of the process is a *new statement and a new principle*.

The flow of energy transformation processes is governed both by a purely quantitative law of the conservation of energy and a law determining the *direction* of processes, i.e. a specific law for their flow.

The establishment of the one-way character of thermodynamic processes leads to very important general consequences. The heat radiated by the sun is the source of all terrestrial processes in operation. All physical processes ultimately are transformed into thermal ones. E.g. our pendulum, if left alone, will swing until its stored mechanical energy has been transferred to the surrounding air as heat energy due to the resistance of the air.

Electrical current will heat wires and heat up electrical light bulbs or start engines whose mechanical energy will be transferred, e.g. to a boring mill. Mechanical energy will be transformed into thermal during the process of boring, the latter will heat the bore and the processed object. In all these processes—mechanical energy of a pendulum, the energy of an electric current in wires, in a bulb, in an engine and in a mill—will ultimately be transformed into heat energy. However, as soon as this transformation had occurred, we would be unable to reverse the process, i.e. to make a pendulum go up to the point B due to the cooling of the air, or to reproduce an electric current due to the heat received by the environment from wires, bulbs or engines.

However, since all processes ultimately transfer into heat energy, and thermal processes, as we saw, proceed in one direction, the main trend in energy transformation is in energy "equalisation".

Indeed, if a hot body is placed in a cooler environment then heat will transfer from the body to the environment until temperatures equalise, i.e. thermal equilibrium is reached.

The heat energy of a hotter body is, as it were, on a higher level characterised by a higher temperature. Energy passed from the higher to the lower level. The thermal levels of the body and the environment were equalised. The total amount of energy

[4]TN: Often called the Kelvin-Planck law, this formulation is due to Planck, Planck (2013), p. 89.

stayed the same but energy, as it were, lost value as it lost its ability for further transformation. It passed into an equilibrium, was equalised or in customary terms, was dissipated.

For this reason, the second law is sometimes called the law of energy dissipation.

The *quality* if not the *quantity* of energy is lost. All thermal processes seek their unavoidable end—thermal equilibrium. This thermal balance is seemingly the death of the world. Different levels are necessary for heat energy to transfer into other types, but the *one-way* direction of thermal processes leads to the inescapable equalisation of levels, and a new difference in levels cannot follow. The motion of energy stops. Energy degenerates or degrades. Stillness of death is the inevitable end of the world predicted by the law of energy dissipation.

We see that the second law of thermodynamics goes much further in its consequences than the first. By determining the processes' direction, it thereby determines their end. However, if processes inevitably proceed to the end their *beginning* is also somewhat mysterious.

If a process ends and is unable to resume by itself, then its beginning (i.e. the occurrence of the initial different levels) which determines its further flow is conceivable only as a creative act of a specific force extraneous to matter, because none of the known natural forces can produce the difference in thermal levels after equilibrium is established. I can certainly warm up the cooled ball and create a difference in thermal levels, then put the ball back into the glass but I can do this only because the equalisation of the thermal levels of the glass and the ball is not yet a global equalisation, and higher thermal levels are still to be found in nature. However, if complete thermal equilibrium were to be established no physical force would be able to bring the world out of it.

These are the conclusions that must arise from the law of energy dissipation according to Clausius. This law indeed divides physical processes by an impassable abyss. While the law of energy conservation connects all natural processes and shows their unity, the law of energy dissipation seems to limit this connection and draws an insurmountable distinction between reversible and irreversible processes.

Clausius's purely phenomenological formulation of the law of energy dissipation is unsatisfactory. This was pointed out by Engels with the utmost clarity and sagacity.[5]

L. Boltzmann's merit is in revealing the essence of the specific distinction between reversible and irreversible processes and in showing how to overcome the above-mentioned difficulties which result from the formulation by Clausius. Let us first look at the essence of the specific character of irreversible processes compared to reversible or mechanical ones.

When kinetic gas theory attempts to give a mechanical interpretation to the law of energy dissipation, it immediately stumbles across the following fundamental difficulty.

[5]TN: *Dialectics of Nature*, Engels (1988), p. 562–563. Engels criticises Clausius for concluding from the second law that the energy existing in the universe is not conserved and therefore must have been created, i.e. an impulse from outside was necessary.

The thermal motion of molecules is a mechanical process. Heat is nothing but the kinetic energy of molecules in flight.

However, we said earlier that every mechanical motion is *reversible. Why then are thermal processes irreversible if they are nothing but the mechanical motion of billions of molecules from the standpoint of the kinetic gas theory of matter?*

When L. Boltzmann formulated his kinetic interpretation of the law of energy dissipation, i.e. the so-called H-theorem, he was immediately reproached for turning to irreversible processes (significantly different from mechanical laws) when examining a gas consisting of molecules. The gas presents a purely mechanical, and therefore, a reversible model and its molecules move under the laws of mechanics. Laws that are significantly different from those governing the motion of individual molecules arise in a gas as a whole.

Why then, based on a purely mechanical gas model, do we arrive at irreversible processes, i.e. completely opposite to the mechanical processes that are always reversible?

Let us hear Comrade Timiryazev's answer to this question.

> Objections to the Boltzmann theorem boil down to the impossibility for the laws of mechanics which reflect strictly reversible processes, to lead to the reflection of an irreversible and one-sided process of transition from any distribution of velocities to the Maxwell distribution [expressed by the one-sided change of H (and its transition to a minimum value)][6] . . . The fact of the matter is that while deriving this theorem, we use probability theory for calculating the number of collisions of this or that type. Therefore, the H theorem should not be regarded as a consequence of *only mechanical equations*.[7]

This is the key.

Laws appearing in gases as a whole aggregate of molecules are specifically different from purely mechanical laws that govern the motions of a single molecule. These laws do not result only from mechanical equations.

What assumptions should be then made *as well as* the laws of mechanics in order to give a kinetic interpretation to the law of energy dissipation?

We imagine a gas as an aggregate of molecules. Molecules suffer a vast number of collisions every second. In order to derive Boltzmann's theorem, *apart from* the assumption of mechanical laws for the motion of molecules, a special assumption of molecular velocities is necessary. Clearly, molecules move at different speeds. However, we assume that molecules with different speeds are evenly distributed in space. In other words, a molecule with a given speed can be located at any point in space with equal probability. Molecules with a given speed are not clustered together and are evenly distributed throughout the entire gas volume.

This modest (on the face of it) statement called the "molecular chaos hypothesis"[8] plays a fundamental part in the entire kinetic gas theory and is essential for it.

---

[6]TN: The phrase in parentheses is from Timiryazev's original and not included by Hessen. It is added for clarity.

[7]TN: This quote is from A.K. Timiryazev, *Kinetic Theory of Matter*, Gosizdat, Moscow-Petrograd, 1923, p. 88.

[8]TN: Also called Stosszahlansatz.

What type of a hypothesis is this? Why do we call it non-mechanical together with Boltzmann's theorem?

Firstly, it is relevant to the gas as a whole and reflects the specific law for the distribution of molecules in a given aggregate. It loses its meaning if applied to an individual molecule. Furthermore, this law for the distribution of molecules inside an aggregate is not contained in the mechanical law of motion for an individual molecule. It arises only from *the aggregate* of these mechanical motions. *An aggregate of mechanical motions of single molecules creates a specific law relevant for the entire aggregate of molecules—a non-mechanical law.*

This is our understanding of the specificity and irreducibility of phenomena and laws observed in gases as a whole to the mechanical laws for gas elements (molecules).

We arrive at the specific distinction between reversible and irreversible processes only thanks to this specific law arising in an aggregate of molecules which is fundamentally different from the elementary laws of motion for individual molecules. We shall see later that the specific nature of this law for the whole will allow us to establish both the distinction between reversible and irreversible processes, and their unity.

Do the mechanical laws governing the motion of single molecules continue to work in the volume of gas? Certainly. Do any *specific supernatural forces* that are unusual for a mechanical system arise in an aggregate of molecules? *No, they do not.*

Is there a specific difference between the law observed in gases as a whole and the laws for gas elements? *Without a doubt.* The molecular chaos is this specific difference.

Does this specificity of the law result from some unusual forces or is it founded on the same mechanical laws for the motion of single molecules?

The specific nature of laws for the whole arises not from some specific forces but from the unification (synthesis) of a vast number of elementary laws into one aggregate. These laws are *fundamentally different.*

The specificity of the law for gases as a whole is known in kinetic gas theory as "the molecular chaos hypothesis" and is caused by the mechanical motion of single molecules; it results from this motion, is completely different from and is irreducible to it.

Comrade Stepanov teaches us, that "scientific biology, applying physical and chemical methods to the study of nature, reveals the *same laws* in living processes as those observed in the domain of dead nature."[9]

We say, no. This premise is applicable neither in biology nor in physics. Physics discovers fundamentally different non-mechanical laws through the use of mechanical methods in kinetic gas theory.

By applying the laws of mechanics to [the study of] kinetic gas theory we arrive at the interpretation of irreversible phenomena through the reversible (mechanical) ones only because we discover such laws in the mechanical gas model consisting of

---

[9]BH: Stepanov, "Dialectical materialism and the Deborin school", p. 61.

a vast number of molecules, that are not present in the mechanical equations of the molecular motion.

A Mechanist might argue that he accepts our specificity and even might remove his reproach for our introduction of some unusual supernatural forces. But why do we talk about the *irreducibility* of these specific laws for the whole to simple laws of its elementary parts? What do we mean and how do we justify this statement?

Should he have the exact equations of motion of all molecules of a given volume of gas and their positions at a given moment in time this would suffice in order to derive our hypothesis of molecular chaos with all its specificity.

Were he able to establish this specific law based on the full knowledge of *only* the molecular equations of motion this would then signify that the above law is *reducible* to these simple laws. Indeed, he cannot do it at present as current knowledge is still infinitely insufficient. But you pose a fundamental question and have maintained more than once that your standpoint does not rely on current scientific knowledge!

Dear reader, let us open a book called *A Philosophical Essay on Probabilities* written by Laplace in 1795.

> Given for one instant an intelligence which could comprehend all the forces by which nature is animated and the respective situation of the beings who compose it—an intelligence sufficiently vast to submit these data to analysis—it would embrace in the same formula the movements of the greatest bodies of the universe and those of the lightest atom; for it, nothing would be uncertain and the future, as the past, would be present to its eyes.[10]

The fundamental formulation of the reducibility problem by Laplace and by the Mechanists is identical. If we were exactly and completely familiar with all laws of motion and positions of elementary particles, we then would have reduced all phenomena to such motion and positions and would have derived all laws for the universe and all states of the universe at any moment in time.

According to Comrade Stepanov the methodological content of the problem would not change in the least if physical and chemical forces were introduced as well as the motion and the positions of particles. We demonstrated in the previous chapter that according to the true "post-chemical" mechanical approach physical and chemical phenomena must be reduced to mechanical ones. However, should we pause halfway through the reduction, as Comrade Stepanov does, and accept that physical and chemical processes are indecomposable and irreducible to mechanics, all the same we repeat: the formulation by the Mechanists and Laplace is identical.

However, while Laplace is a consistent Mechanist and understands forces and laws as mechanical, our Mechanists imply physical and chemical forces.

Ergo, we even accept that physical and chemical phenomena are irreducible to mechanics, and all other phenomena are reducible to physical and chemical laws. Certainly, this is not consistent: if Mechanists believe that all phenomena are *fundamentally* reducible to physical and chemical ones then why are not physical phenomena reducible, e.g. to simple mechanical ones?

What does this formulation mean?

---

[10]BH. Russian translation by Vlasov, p. 12. TN: English translation from Laplace (1902), p. 3.

This formulation is accepted by every materialist if Mechanists thus want to say that nature as a whole contains all natural laws, that there are no other forces apart from the ones at the foundation of elementary natural processes, that all laws for animate nature are essentially linked only to the laws governing inanimate nature, and there are no laws and forces above and beyond nature and can be none—in short, if the reducibility is understood as a genetic connection between higher and lower forms in the development process. It would be useful if Comrade Stepanov and other Mechanists, instead of abuse, were to quote a single line to prove that we[11] accept the existence of supernatural forces and laws unconnected with nature and not embedded in it.

Undoubtedly, this materialistic premise is enclosed in the Laplace formulation. Laplace understood this well. This is why when Napoleon asked why in his celestial mechanics he left no room for God he answered, "I saw no need for this hypothesis."[12] Laplace's formula, however, contains something more than a simple materialistic premise. It is a purely methodological statement that the direction of scientific development is in the decomposition of all phenomena into elementary ones and in the reduction of all laws to the law for the elements. The specific nature of laws for the whole in this formulation arises from our lack of knowledge.

Specificity is a subjective category, and the scientific task is in the reduction of this specific law for an aggregate to simple laws for its parts. Specificity is a knot, and the task of science, according to Comrade Stepanov, is to untie it.

It is not the materialistic aspect of the formula of reduction we are opposed to but this methodological premise.

Whether a premise is correct or incorrect is decided by scientific practice.

What do scientific achievements teach us and in particular, the above-mentioned example of the kinetic interpretation of the law of entropy? What direction did science take? Was it the path of reduction or the path of the establishment of a specific (irreducible) law for the whole *as well as* the law for the elements?

It took the second path.

Certainly, the molecular chaos hypothesis is inherent in the motions of individual molecules. It does not appear as a deus ex machina. It is conditioned by these mechanical processes. But if science followed the approach of the derivation of laws for a gas as a whole from the motions of individual molecules, we would without a doubt have had no kinetic gas theory.

The image of the omniscient intelligence painted by Laplace is nothing but an abstraction of the infinite process of knowledge.

This intelligence must possess an exhaustive knowledge of all laws and of the structure of nature (the positions of elementary particles). Both are inaccessible to real knowledge as it would mean that nature is fully known and decomposed into the final indivisible absolutely elementary particles and laws.

---

[11]TN: i.e. the Deborin group.

[12]TN: The alleged reply of Laplace to Napoleon is not in his writings. What he actually said is now disputed.

Perhaps, L. Boltzmann described best the abstract and metaphysical character of the mechanical approach and of the "reducibility" principle as the basis for scientific research:

> It was now attempted to prove a priori that every change, even if apparently qualitative, must be reducible to a motion of the smallest parts, motion[13] being the only process in which the object moved remains always the same. All such metaphysical reasons seem to me to be insufficient. Of course we cannot avoid forming the concept of motion. If therefore all apparently qualitative changes were representable by the picture of motions or changes of arrangement of smallest parts, this would lead to an especially simple explanation of nature. In that case nature would appear to us at its most comprehensible, but we cannot compel her to this, we must leave open a possibility that this will not do and that we need in addition other pictures of other changes; understandably, *it is precisely the more recent developments of physics that have made it prudent to allow for this possibility.*[14]

Boltzmann rightly emphasizes the metaphysical nature of the mechanical concept. Indeed, in order to regard the reduction of every natural phenomenon to mechanical laws or to the motion of electrons at the forefront it is necessary once and for all a priori to postulate that mechanical motion is the primary basis of the motion of matter. Here mechanical motion is understood as both the motion of discrete particles and the motion of continuous media. By supposing the existence of electromagnetic phenomena that are *irreducible to any type* of mechanical motion we essentially renounce the mechanistic worldview regardless of any verbal twist about the distinction between mechanical and mechanistic. This is being used to gloss over the assumption of new specific forms of motion of matter which are qualitatively different from and irreducible to mechanical ones.

We can repeat after Boltzmann that the necessity of such [mechanical] assumptions is not proven; moreover, physics in its development follows a different path. We shall come back to this in the chapter on electronic theory.

And yet, a Mechanist will argue, do you deny the necessity of the electronic understanding of chemistry or not? Do you not consider the attempted application of electronic theory in chemistry a step forward?

Undoubtedly, we do. Any establishment of connections between different forms of material motion always marks scientific progress. Electronic processes are one form of material motion. Chemical laws are a different one. We do not oppose the reduction if it means connection between different types of laws of material motion.

However, Comrade Stepanov poses the problem differently:

> The known combination of physical and chemical processes occurring in inanimate matter once led to the nodal line, to the jump, to life or to a new quality. But this does not at all signify that some processes unknown in physics or chemistry unfold in an organism and that science should cast aside an utterly vain hope to reduce life phenomena, however original and complex, to relatively simple ones scientifically observed in inanimate nature.[15]

---

[13]BH: Boltzmann naturally means mechanical motion.

[14]BH: Speech, "On the principles of mechanics" Populare Schriften. p. 325. TN: English translation from Boltzmann (1974), pp. 142–143. Italics added by Hessen.

[15]BH: Stepanov, "Dialectical materialism and the Deborin school", pp. 35–36.

To start with, what does it mean to reduce more complex (biological) phenomena to "relatively simple" ones? "Relatively simple" in such a formulation has, in a sense, a conditional character. E.g., we think that simple mechanical motion is the simplest and the most graphic one. Boltzmann is right to emphasize that the mechanical world view is the simplest (from our standpoint) explanation of nature. Yet, nature does not care whether we find it easy to know this or that law, whether our differential equations are simple, and whether we can integrate them or not. Therefore, objectively we cannot say whether the mechanical motion of a neutral material point is simpler than the motion of an electron. Oliver Lodge beautifully expressed this in his speech on Kelvin on 19 April 1928.

> Modern physics aims at simplifying the complex by the aid of relativity and quanta, but it has raised difficulties where previously we detected none, and has made simple things complex. . . In the nineteenth century everything was reduced to mechanics; now the very motion of matter itself is in need of explanation.[16]

The main task of science is not a simple reduction of the complicated to the relatively simple. It is the connection between different forms of material motion and the study of *significant* laws of motion of matter (naturally, we understand motion according to Engels as a general change).

In his speech "The Aim and Progress of Physical Science" Helmholtz says, "To find the law by which they are regulated is to understand phenomena".[17]

And he is quite right. A consistent Mechanist, Helmholtz believed that everything could be reduced to mechanics. At the same time, he did not formulate the scientific task as that of reduction.

We might be asked, "Is there no objective distinction between the simple and the complex? Do you really claim that biological laws cannot be objectively called more complicated than physical or chemical ones?"

The difference between the simple and the complex can be established only if we assume the standpoint of the *development* of the complicated from the simple or an aggregate from its elements. Yet this very standpoint of the development is blurred by the reduction formula.

The world is moving matter. The motion of matter in the process of development gives rise to yet more new original forms. To study nature means to establish significant laws and connections between different forms of motion and *to establish the law of their development*. Speaking of simple and complicated forms makes sense only from the standpoint of the development. Whereas the aspect of development for Comrade Stepanov fades into the background.

---

[16]BH: *Nature*, No. 3073, 22 September 1928, p. 430. TN: The quotation is taken from a comment article "The Revolution in Physics", *Nature*, No. 3073, Vol. 122, September 22 1928, pp. 429–431. It refers to Oliver Lodge's Kelvin lecture to the Institution of Electrical Engineers (with a reference Journ. Inst. Elec. Eng., Vol. 66, 1928, p. 100), but is not a quotation from this lecture as Hessen mistakenly assumes.

[17]TN: Helmholtz (1995), p. 208.

A chemical element is built of electrons. Living matter is built of chemical elements. The development process in nature proceeds from non-organic to organic matter.

Can we then say that a chemical element is more complex than an electron, and that living matter is more complex than chemical elements?

Only in the sense that new specific laws that are typical for a given formation and are not included in its elements arise in living matter (a later link in the development of non-organic matter) compared to a chemical element. Therefore, the scientific task is to study specific laws typical for a new formation (an atom from electrons), and to establish a connection between this new specific law and the elementary laws that underlie it.

It is important and necessary to study the connection between physical and chemical laws and the phenomena in organic matter.

It is important and necessary to study the connection between electrons and chemical elements.

It is equally important to study specific laws of living phenomena and specific properties of an atom because the new formation is a later link in the chain of development. This link is always richer in content and presents new laws that are specifically different from the laws of its elements. The example of kinetic gas theory showed us that the law for a gas as a whole (the molecular chaos) presents a specific law compared to the mechanical motion of molecules.

Only the establishment of a connection between the specific law of the whole (volume of gas) and the mechanical motion of its elements (molecules) results in kinetic gas theory with its new specific laws, i.e. statistical laws. Mechanics will never abolish kinetic gas theory and physics in the same way as biology will never become a substitute for sociology.

To summarise.

There are two types of relationships between a whole (an aggregate) and its parts or between the simple and the compound in nature. The whole might be a simple arithmetical sum of its components. Properties and laws for the whole and for the parts are the same. A kilo (in weight) is nothing but an arithmetical sum of 1,000 g. No matter how many times we add 1,000 g (in weight) we shall never get any new properties.

A kilo is entirely decomposed into a sum of grams, i.e. is "reduced" to 1,000 g.

We call these properties of aggregates additive properties.

An aggregate can also possess properties that are not included in its elements. If two different metals are brought into contact we will get a contact potential difference. This potential difference as such is not an attribute of any metal but is a new property of the aggregate that is not included in its elements. It presents a new specific formation. These properties of an aggregate are conditioned by the laws of the elements but appear only in the aggregate as a whole and disappear when it is decomposed into its parts. They result from the synthesis of elements and are called non-additive properties of an aggregate.

Should only additive properties exist in nature we would rightly speak of the process of reduction. A kilo is indeed reducible to a sum of grams. A kilowatt is reducible to 1000 W.

However, additive properties of an aggregate cannot result in a *process of development* because this process is characterised precisely by the *appearance of new properties*. Since dialectics is a theory of development, our main task is to study new properties of the higher forms and to establish the connection between these new and the elementary forms. In other words, to study the forms of motion of matter in their development.

This is why the task is not to reduce specific laws of a whole to elementary laws of the parts. It is to study *both* in their mutual connection and development. We cannot reduce or dilute a higher form of motion to the sum of lower ones. The specificity of a higher form of motion is precisely in the synthesis of the lower forms and not their sum.

Comrade Stepanov demands, "Be so kind as to explain this mysterious synthesis". This is easy enough because Engels gave this definition of synthesis:

> Simple and compound. Categories which even in organic nature likewise lose their meaning and become inapplicable. An animal is expressed neither by its mechanical composition from bones, blood, gristle, muscles, tissues, etc., nor by its chemical composition from the elements.[18]

An organism is a synthesis and not a sum of tissues, bones etc. A living tissue is a synthesis of chemical elements and not their sum. A chemical element is a synthesis of electrons and not their simple assembly. We call it a synthesis because new specific properties and laws arise in a higher (from the standpoint of the development) form that are not present in its elements. It is the emergence of these new specific properties and laws that is significant for a synthetic union of elements as opposed to a simple sum.

To decompose a synthesis into the main elements means to destroy these specific properties and laws. This is why analysis by itself is not sufficient. Chemistry perfectly knows the analytical composition of live matter (protein) but it cannot create a protein just on the basis of this knowledge. Therefore, synthetic chemistry exists alongside analytical chemistry. The former studies the specific laws of the synthesised substances from the point of view of the methods and laws of their emergence.

Comrade Stepanov is absolutely wrong in saying, "When chemistry will study (*analytically, as no other method exists*[19]) the protein structure it will be able to synthesise it. How else would it (natural science—BH) have studied metabolism in a living organism and would have arrived at scientific agriculture?"

One of our most significant disagreements is ignoring synthetic methods of studying phenomena.

The point is that methods of synthetic study or synthesis exist as well as analytical methods.

---

[18]BH: Archive, II, p. 117. TN: *Dialectics of Nature*, Engels (1988), pp. 494–495. Engels cites Hegel (2010), p. 194.

[19]BH. Our italics. Stepanov,"Dialectical materialism and the Deborin school", p. 47.

Comrade Stepanov says, "Empty phraseology. No other methods exist or can exist apart from the analytical methods."

Let us turn to Klimenti Arkadievich Timiryazev[20] for the resolution of our dispute.

Here is what he writes about analysis and synthesis in his remarkable essay on Marcellin Berthelot:

> ... Even the most outstanding chemists were infected with the vitalist point of view borrowed from the medical scientists up until the mid-nineteenth century because they were aware of their own powerlessness before one of the two main tasks of their science. This problem arose as soon as they entered the incomparably more complicated realm of living matter. The two tasks were analysis and synthesis. There was no significant difference between these two processes in the realm of inanimate matter. Perhaps one can say that the synthesis of bodies was known before from analysis, and the main task was the analysis. Lavoisier particularly insisted that a chemist "divides and subdivides and subdivides again" precisely because knowledge of synthesis in inanimate nature preceded analysis in the majority of cases.
>
> ... The picture completely changes with the transition to the world of living beings, i.e. to organisms and their components that are called organic. This is due to the fact that innumerable experiments convinced [the scientists] that they are encountered only in organisms and that only organisms possess the mystery of their creation. Already Lavoisier by dividing and subdividing organic compounds decomposed them into the same elements which made up non-organic bodies. But neither he nor anyone else until Berthelot attempted to create organic matter from elements. Analysis alone ruled in the realm of organic chemistry, at the same time synthesis was considered a mystery of life or a result of the acts of a mysterious living force. Normal physical forces were insufficient for this task – this was the slogan of the ruling idea of vitalism.
>
> ... The first attempt was to resolve the problem of the best approximation [sic] to the composition of natural fats and to synthetically create them from their nearest components.[21]

We see that Comrade Stepanov's formulation does not at all cover the entire complexity of the problems facing natural science and chemistry in particular.

Undoubtedly analysis is a powerful method of research. However, this does not mean that analysis is the only "method because none other exists".

Analysis is the decomposition of an aggregate into its components. A study of the parts in isolation is one aspect of the cognition of nature. Synthesis is the study of the way to develop the whole from its parts and to find the specific laws for the whole as a new formation. It is an essential second aspect of cognition.

To paraphrase Comrade Stepanov:

To believe that the task of scientific knowledge of nature is only analysis, dissection into components and the study of the laws of these isolated parts (for isolated they are, because their study in their mutual connection would be a synthetic study) is to reject a complete knowledge of nature. It would limit the task to the study of the additive properties that are preserved upon the disintegration of the aggregate.

The true task of scientific knowledge, however, is to study the significant laws of the forms of material motion in their mutual connection, interaction and development.

---

[20]TN: Kliment Arkadievich Timiryazev, 1843–1920. The founder of the Russian scientific school of plant physiology. His son was Arkadii Klimentievich Timiryazev (see Appendix, Introduction and Chap. 3).

[21]BH: Kliment Timiryazev, Collection "Science and democracy", pp. 200–201.

References

Boltzmann, L. (1974). *Theoretical physics and philosophical problems*. Dordrecht: Reidel.
Clausius, R. (1879). *The mechanical theory of heat* (W.R. Browne, Trans.). London: Macmillan and Co.
Engels, F. (1988). *Marx engels collected works* (Vol. 25). Moscow: Progress Publishers.
Hegel, G. (2010). *Encyclopedia of the philosophical sciences in basic outline, part i: Science of logic* (K. Brinkmann & D.O. Dahlstrom, Trans.). Cambridge: Cambridge University Press.
Helmholtz, H. (1995). *Popular and philosophical essays*, D. Cahan (Ed.), Chicago: University of Chicago Press.
Laplace, P. S. (1902). *A philosophical essay on probabilities*. London: Chapman and Hall.
Planck, M. (2013). *Treatise on thermodynamics* (A. Ogg, original edition Longmans Green, London (1903). Trans.). New York: Dover Reprint.

Chapter 8
Mechanical Materialism and Modern Physics (Section 3)

Boris Hessen

III. Statistical method in physics and the problem of chance and necessity.

Let us go back to the mechanical interpretation of the law of energy dissipation and examine it from a new point of view. Earlier we have emphasized the specific distinction between reducible (mechanical) and irreducible processes. We shall now look at their unity. The establishment of the relationship between the specificity and unity of the reducible and irreducible processes enables us to approach a new type of law, i.e. the statistical one. The problem of statistical laws will allow us to understand the importance of interpreting the concept of chance as an objective category.

Thus, before turning to the statistical interpretation of the law of the dissipation of energy we should make a rather detailed introduction to the character of the two main types of physical laws, statistical and dynamical, and also find out the connection between probability and chance.

Maxwell with his typical foresight posed the question on the new type of law brought about by the emerging kinetic gas theory.

In one of his less known philosophical essays[1] he says,

---

[1] BH: Maxwell looks into the problem of necessity and chance applied to physics in a small article published by Campbell and Garnett—Maxwell's paper was given to a philosophical group at Cambridge (Club of Seniors). The article is titled "Does the progress of Physical Science tend to give any advantage to the opinion of Necessity (or Determinism) over that of the Contingency of Events and the Freedom of the Will?" TN: See Campbell and Garnett (1882), pp. 209–213. See also Chap. 4, p. 57, n. 45.

TN: Translated from *Pod Znamenem Marksizma* (*Under the Banner of Marxism*), 1928, Nos 7–8.

B. Hessen (✉)
Abingdon, UK
e-mail: chris.talbot@protonmail.com

C. Talbot and O. Pattison (eds.), *Boris Hessen: Physics and Philosophy in the Soviet Union, 1927–1931*, History of Physics,
https://doi.org/10.1007/978-3-030-70045-4_8

> But I think the most important effect of molecular science on our way of thinking will be that it forces on our attention the distinction between two kinds of knowledge, which we may call for convenience the Dynamical and Statistical.[2]

One of the typical key features of the distinction between statistical and dynamical laws is the fact that a dynamical law concerns a *single, individual phenomenon* while a statistical law is for an *aggregate* of individuals or phenomena.

A dynamical law is not suitable for the study of the laws of a collective. But a different type of law, a statistical one, is.

The opposition between statistical and dynamical laws is the opposition between macro- and micro-worlds, or between an individual and a collective, in the same way that reversible and irreversible processes oppose each other.

We shall look at the most crucial features of statistical laws.

An opposition between an individual[3] and a collective is mainly that between a whole and a part. A collective differs from a simple sum of individuals (parts) precisely because it is a new formation with new properties that are not inherent in single individuals and appear only in a collective. Therefore, a collective *consists* of a certain number of individuals but is not *reducible* to single individuals. It is particularly important to note for our purposes that the properties of a collective are not decomposable into a simple sum of properties of its parts.

In the study of dynamical laws, one considers properties of single individuals and, therefore, these laws are inapplicable to studying collectives. When studying a collective we are interested in its properties and laws and not a single individual itself and its properties.

While individual phenomena cannot be studied using statistical laws because these laws are simply inapplicable to individuals, the applicability of dynamical laws to the study of aggregates or collectives is not so simple. Any aggregate consists of individuals. Therefore, it seems on the face of it that dynamical laws are fully applicable to the study of an aggregate. However, the situation is more complicated.

Two approaches, i.e. subjective and objective, are possible regarding statistical laws.

An aggregate consists of a vast number of individuals. Each individual behaviour is univocally determined by a dynamical law.

An aggregate can be studied as a sum of a vast number of dynamical laws. Such an approach is possible but extremely complicated. Thus, we turn to a statistical law which, notwithstanding it being second-rate knowledge, can successfully complement the lack of our knowledge and can partly overcome the extraordinary difficulties arising in studying an aggregate of a vast number of individuals.

---

[2]TN: Original English from Campbell and Garnett (1882), p. 210.

[3]BH: By "individual" hereinafter we mean objects studied by physics, i.e. a molecule, an atom or an electron. Collectives are the bodies that are macroscopic in relation to them. Broadly speaking, a collective may be considered as an individual if it is included in a larger collective. A molecule is an individual in relation to a gas volume but is a collective in relation to its constituent electrons.

The above reasoning is easily recognised as a typical argument of a *subjective* approach to statistical laws. In this sense statistical laws are a consequence of the limitation of our cognitive abilities.

An *objective* approach to statistical laws means that the raison d'être of these laws is not in the limitation of our knowledge. It is in the particular characteristic structure of the objects studied with its aid, i.e. aggregates.

An aggregate studied by the statistical method is examined as a *whole*.[4]

Although an aggregate consists of individual elements it does not decompose into separate elements *in the process of our study*. It is studied as a whole, i.e. synthetically. Therefore, statistical laws are applicable to an aggregate as a *whole* and make no sense if applied to individual elements.

But what does it mean that statistical laws work for an aggregate as a whole but are not applicable to its individual elements? In other words, what is the relationship between dynamical laws that govern constituent individual elements of an aggregate and statistical laws applicable only to an aggregate as a whole?

Non-additive properties are typical for an aggregate as a whole and only as a whole. They are not virtually included in its constituent individuals. They manifest themselves only in a whole and are *qualitatively* different from the properties of individuals.

It is these non-additive properties that are noted and studied by statistical laws. It is clear, therefore, that statistical laws in their very essence cannot apply to single individuals that constitute an aggregate. This is their *characteristic feature* and not their *deficiency* because they study the properties that manifest themselves only in a whole and *do not exist* in individual members.

We consider the presence of non-additive properties in an aggregate characteristic of its objective structure, and thereby, we believe in the objective character of statistical laws.

In this sense the relationship between statistical and dynamical laws is between laws for a whole and for its parts. Dynamical laws remain individual laws. Yet they are not sufficient for the study of the law for the whole because the whole has both additive and non-additive properties.

Statistical laws do not negate and do not oppose dynamical laws. They are necessary and valid in their field—while dynamical ones are in theirs.

M. Planck[5] rightly believes that a dynamical law is a condition for the occurrence of a statistical law. However, this does not mean that a statistical law is reducible to a dynamical law without a remainder. This is correct only if the whole is identical to the sum of its parts.

If this identity is absent (and the presence of non-additive properties in an aggregate affirms this absence) then a statistical law genetically emerges from dynamical laws in the same way as a whole arises from a part. However, a statistical law is not composed of and cannot be decomposed into dynamical laws but is a qualitatively

---

[4]TN: Russian: *kak tseloe*.

[5]BH: Essay "Dynamical and statistical laws" in the collection "Essays on physics". TN: English translation in Planck (1960), pp. 56–68.

new formation, intrinsic only to a whole and not to its parts. Therefore, statistical laws are not second-rate knowledge compared to dynamical laws. They are a completely equal method of cognition conditioned by the peculiarity of the objective structure of the studied objects.

The concept of probability is introduced in order to express statistical laws.

This concept is inextricably connected to the concept of chance and therefore, obtains a subjective or an objective character depending on our interpretation of chance as a subjective or an objective category. The same goes for the concept of a statistical law. We see now that our discussion about the objectivity of chance is of paramount importance in physics.

Before examining the problem of chance let us clarify the connection between the concepts of probability and chance.

Already Laplace pointed to the connection between these two concepts,

> All events, even those which on account of their insignificance do not seem to follow the great laws of nature, are a result of it just as necessary as the revolutions of the sun. In ignorance of the ties which unite such events to the entire system of the universe, they have been made to depend upon final causes or upon hazard,[6] according as they occur and are repeated with regularity, or appear without regard to order; but these imaginary causes have gradually receded with the widening bounds of knowledge and disappear entirely before sound philosophy, which sees in them only the expression of our ignorance of the true causes.
>
> ... The curve described by a simple molecule of air or vapor is regulated in a manner just as certain as the planetary orbits; the only difference between them is that which comes from our ignorance. Probability is relative, in part to this ignorance, in part to our knowledge.[7]

Chance does not exist for Laplace's "omnipresent Intelligence" for whom all knowledge is certain. However, for our limited minds chance phenomena exist. Therefore, probabilistic knowledge emerges together with certain knowledge. I toss a coin. The process of tossing is an extremely complex aggregate of phenomena. We are unable to exactly calculate how the coin falls, heads or tails. This is why we say that heads or tails are due to chance. We say so because we do not know which will fall. We know for certain that one of them will. However, which of them falls when we toss the coin this time is only probable and not certain.

Probability theory determines the value of the probability of a phenomenon. If our coin is a regular one (fully symmetrical) we say that the probability of heads is equal to a half. However, this does not mean that if tails fall this time then heads will be next time. Probability equal to a half means that when we toss a *large number* of times and count the number of heads and tails they are distributed almost equally; and the number of heads (or tails) is the closer to a half of all tosses the more we toss the coin.

Thereby, probability expresses the law of distribution of heads and tails *inside the entire aggregate of tosses* and says nothing about a single toss.

---

[6]TN: The English translation uses the word "hazard", but "chance" is the usual term used in physics.

[7]BH: "A Philosophical Essay on Probabilities", translated by Vlasov, p. 11. TN: English translation from Laplace (1902), pp. 3 and 6.

Even if we know the probability of heads we know nothing about a single toss's result.

Probability expresses a statistical law, i.e. a law applied to the entire aggregate of tosses as a whole. This is why we say that probability is a method of expressing statistical laws.

The concept of probability is closely connected to the concept of chance. We saw earlier how Laplace defines this connection.

He gives a purely subjective interpretation of chance. Chance is a measure of our ignorance. Therefore, in this sense probability is also a subjective concept. And with this concept statistical laws are also a subjective category. Knowledge of statistical laws is second-rate knowledge compared to dynamical laws.

We see how very important the resolution of the problem of chance is in physics. It conditions our interpretation of probability as well as of statistical laws.

The fierce attacks on Dialecticians who defended the point of view that chance is an objective category are well known.

It came to the point of accusing Dialecticians of defending the viewpoint of acausality.

The basis for this absurd accusation is easy to understand. We saw earlier that Laplace (he was certainly not the first and not alone) considered chance to be that whose causes were unknown to us. On this basis the conclusion is made that chance is that whose causes are unknown. It does not contradict the general premise of determinism in the subjective sense. "*Chance is only the measure of our ignorance*".[8] But if the concept of chance is made objective then a subjective absence of causes (ignorance) is transformed into an objective absence of causes, i.e. into acausality.

This reasoning is not accidental. However, it is simply a measure of ignorance.

The dialectical view allows us to attribute an objective meaning to chance and at the same time to preserve the deterministic point of view. However, the mechanical concept of chance certainly is inadequate.

Let us look at a series of 10,000 coin tosses. From the point of view of a subjective approach to probability the number of heads and tails and the way they alternate are accidental because we are unable to predict their sequence *due to our ignorance*. From the dialectical point of view the order of heads and tails is accidental because it is not reflected in the main statistical law. According to this law the number of heads approximates the number of tails with the increase of the number of elements in the aggregate.

A single toss is accidental in relation to the law that governs the entire aggregate. This certainly does not mean that no individual law exists for each element. However, although such a law for an element (a toss of a coin) is the basis of the law for the whole, an individual cause of this law has no impact on the behaviour of the entire collective.

---

[8]BH: An expression due to Poincaré. TN: See Poincaré (2003), p. 65.

Let us use the image of a statistical law suggested by Quetelet[9] in order to better illustrate our reasoning.

Let us draw a circle in chalk on a blackboard. This circle is made of a vast number of miniscule particles of chalk that are stuck to the board. If seen through a microscope a part of this circle consists of a chaos of particles that have nothing to do with a circle. If the circle is seen as a whole, e.g. from a distance, then this chaos of individual particles makes a certain geometrical line. An aggregate of laws for the chalk particles' positions makes up a completely new law, i.e. a circle. Positions of individual particles vis-à-vis the circle are accidental. The changing of a particle's position will not affect the law for the entire aggregate of particles of the circular line.

The laws governing the motion of a single molecule are conserved and continue to act in a gas volume that we examine as a collective of molecules. But the laws of this collective embrace the entire aggregate of molecules. The change in motion of a single molecule does not affect the law for the whole. The motion of a single molecule is an accidental process in relation to this law for the whole.

This is what we mean by the objectivity of chance. If all chains of causation were equal, i.e. if according to Engels, "a particular pea-pod contains five peas and not four or six, that a particular dog's tail is five inches long and not a whit longer or shorter, that this year a particular clover flower was fertilised by a bee and another not, and indeed by precisely one particular bee and at a particular time, . . . that last night I was bitten by a flea at four o'clock in the morning, and not at three or five o'clock, and on the right shoulder and not on the left calf—these are all facts which have been produced by an irrevocable concatenation of cause and effect, by an unshatterable necessity of such a nature indeed that the gaseous sphere, from which the solar system was derived, was already so constituted that these events had to happen thus and not otherwise,"[10] then the study of nature would have been impossible.

The fact that we can study nature and empirically distinguish significant from insignificant laws, the accidental from the necessary, proves that causal chains are not all equal, and therefore, the objectivity of chance.

One has only to wonder that L.I. Axelrod in her last book[11] reiterates the objective point of view on chance. One of her subtitles is "Against Scholastics". Comrade Stepanov prudently completely avoids the problem of chance and necessity in his book.

---

[9]TN: Gross (1911), p. 114, gives the quotation: "If you draw a circle on the blackboard with thick chalk, and study its outline closely in small sections, you will find the coarsest irregularities; but if you step far back and study the circle as a whole, its regular, perfect form becomes quite distinct." There is no reference to Quetelet's original. Another version, but with reference to points rather than chalk, can be found in Quetelet (1842), p. 5 (also https://openlibrary.org/books/OL23350005M/A_treatise_on_man_and_the_development_of_his_faculties).

[10]TN: Engels (1988), p. 499. (See also Chap. 4, p. 54, notes 34–35).

[11]TN: Liubov Axelrod, *In Defence of Dialectical Materialism*, Moscow-Leningrad, 1928 (in Russian). Hessen mistakenly refers to the subtitle as "Against Neo-Scholastics.".

We shall now see which point of view is justified by the development of modern natural science.

L. I. Axelrod is quite right if one turns to the time of Laplace. Laplace, Poisson, Bertrand and other classics of the probability theory of the eighteenth and nineteenth century indeed believed chance and probability to be subjective categories.

However, chance is seen in a drastically new light already, starting with Cournot.

He said, "Mathematical probability becomes a measure of physical possibility, and one expression can be used instead of the other. The advantage of this approach is that it emphasizes the existence of (an objective) relationship between things; this relationship does not depend on our changeable method of judgement and assessment. This relationship is in the things themselves."[12]

Cournot's objective approach to probability theory results in a definition of chance that is very close to Plekhanov's definition.

Chance means "events occurring as a result of a combination or an encounter (rencontre) of events which belong to two independent chains of events."[13]

Poincaré in his "Science and Method" criticises Laplace's subjective concept of chance. His definition of chance is close to those by Cournot and Plekhanov.[14]

Notably, the formulation by Hegel (accepted by Engels) is deeper and methodically more effective because it establishes the difference between the essential or necessary law and the inessential or accidental one. It is easy to show that it includes Plekhanov's formulation.

Now then, the resolution of the question of chance determines our attitude to probability theory and to statistical laws. As statistical laws gain dominant importance in modern physics it is clear that the resolution of the question of chance and necessity has a great methodological significance for physics [as a whole.]

M. Smoluchowski particularly sharply emphasizes the growing significance of the statistical method and the need for the clarification for the methodological fundamentals of the concept of probability and chance.[15]

---

[12]BH. Cournot, Essai sur les fondaments de nos connaissances etc., vol. 1., p. 62. TN: Translated from the Russian. French original, Cournot (1851), p. 62: "La probabilité mathématique devient alors la mesure de la possibilité physique, et l'une de ces expressions peut être prisé pour l'autre. L'avantage de celle-ci, c'est d'indiquer nettement l'existence d'un rapport qui ne tient pas à notre manière de juger et d'apprècier, variable d'individual à l'autre, mais qui subsiste entre les choses mêmes . . .".

[13]TN: Translated from the Russian. French original, Cournot (1851), p. 51: "Les événements amenés par la combinaison ou la rencontre d'autres événements qui appartiennent à des series indépendantes les unes des autres . . .".

[14]BH: It will certainly be retorted that Poincaré was a Machist. We shall not enter into a discussion whether Poincaré's criticism of the subjective interpretation of chance is connected with his Machism or not but will turn to the views by Smoluchowski, who was not a Machist. TN: Poincaré (2003), pp. 64–92.

[15]BH: Smoluchowski (1918). Russian translation in *Uspekhi Fizicheskikh Nauk*, No. 7, p. 5 and in *Pod Znamenem Marksizma* (abridged translation), No. 9, 1927.

> After a transitory period of stagnation probability theory has gained a fundamental importance in physics (as a result of the ultimate victory of the atomistic approach) and remains to this day the most important research tool in the field of modern theories of matter, electronics, radioactivity and radiation theory. . . .
>
> In spite of the enormous extension of the application area of probability theory, the exact analysis of its underlying concepts has made but little progress.
>
> To this day it is true that no other mathematical discipline rests on such an unclear and unstable foundation as probability theory. Thus, the questions of subjectivity or objectivity of the probability concept and of the definition of the concept of chance are answered by different authors in diametrically opposite ways. . . .
>
> This study . . . emphatically puts in the right light and properly interprets the objective aspect of the concept of probability hitherto all-too-neglected.[16]

The essay emphatically propels to the forefront and properly interprets the objective aspect of the concept of probability that has been almost neglected until now.

We see that Smoluchowski particularly emphasizes the significance of highlighting the objectivity of probability and consequently, of chance.

He continues,

> I am well aware that this concept of chance stands in contrast to the generally accepted understanding, whereby its essential aspect is the partial ignorance of causes. So may the following be said as attestation of our assertion: probability theory's application in kinetic gas theory would retain its legitimacy even if we knew the exact constitution of molecules and their initial positions, and were able to follow their motion with mathematical precision for all times."[17]
>
> I think that it would also be an extremely important result for a philosopher if it can be demonstrated, albeit in a narrow field of physics, that the concept of probability possesses a strictly objective meaning in its usual interpretation as a regular frequency of accidental events, and that the concept and the genesis of chance can be determined with precision while remaining firmly rooted in determinism.[18]

---

[16]TN: Smoluchowski (1918), p. 253. English translation from the Russian. German original: ". . . hat die Wahrscheinlichkeltsrechnung, nach einer vorübergehenden Periode der Stagnation, infolge des schließlichen Sieges der atomistisehen Anschauungsweise eine für die Physik ganz grundlegende Bedeutung gewonnen und bildet heute das wichtigste Werkzeug bei Forschungen auf dem Gebiete der modernen Theorien der Materie, der Elektronik, Radioaktivität und Strahlungstheorie. . . . Trotz dieser enormen Ausdehnung des Anwendungsbereiches der *Wahrscheinlichkeitsrechnung* hat *die exakte Analyse der ihr zugrunde liegenden Begriffe* nur geringe Fortschritte gemacht; es gilt wohl noch heute der Satz, daß keine zweite mathematische Disziplin auf so unklaren und schwankenden Grundlagen aufgebaut ist. So werden die Grundfragen nach der Subjektivität oder Objektivität des Wahrscheinlichkeitsbegriffes, nach der Definition der Zufälligkeit usw. von verschiedenen Autoren in diametral entgegengesetzter Weise beantwortet. . . Im übrigen bezweckt dieselbe selbstverständlich keineswegs eine allseitige und endgültige Aufklärung des ganzen damit zusammenhängenden Komplexes philosophischer Fragen, sondern will nur eine Anregung zu weiteren Untersuchungen in einer bestimmten Richtung geben, indem einige *Leitgedanken* hervorgehoben werden, welche die bisher allzusehr vernachlässigte *objektive Seite des Wahrscheinlichkeitsbegriffes ins rechte Licht setzen* sollen." (Italics in original, p. 253).

[17]TN: Smoluchowski (1918), p. 254. English translation from the Russian. The same quote is given in Chap. 4, p. 56, n. 43, with the German original.

[18]TN: Smoluchowski (1918), p. 262. English translation from the Russian. The same quote is given in Chap. 4, pp. 56–57, n. 44, with the German original.

The concept of probability can be made precise and scientific only if the concept of chance and probability is interpreted objectively. Extensive literature criticising the fundamentals of probability theory mostly targets the subjective interpretation of probability and chance by Laplace which includes an obvious *petitio principii.* The latest works by von Mises criticise the subjective interpretation of the concept of probability and attempt to justify probability theory on the basis of the objective definition of probability. These works are recognised by physicists in a recently published volume of *Handbuch der Physik*, the most prestigious encyclopaedia of physics, where probability theory is explained on the basis of the objective definition of the probability concept.

And now, following the works by Cournot, Smoluchowski, von Mises and others, and after an exhaustive philosophical substantiation of the question by Hegel and Engels, chance is again pronounced [to be] a subjective category and we are dragged back to Laplace's conception. This is called the struggle against the "neo-scholastics". Our severe critics might be excused only because all recent developments in physics and mathematics are a sealed book to them.

After clarifying the problem of the connection between chance, probability and statistical laws it is easy to establish the connection between the reversible (mechanical) and irreversible (thermal) processes and the essence of the statistical interpretation of the law of energy dissipation.

A body's thermal state can be described by its temperature. Temperature expresses the *degree of a body's heat*. The concept of temperature is unconnected to any hypothesis of the structure of a body. This concept is an empirically and directly observable value (e.g. the position of mercury in a thermometer) that determines the body's thermal state. The law of energy dissipation in the above formulation (by Clausius) is based on the definition of temperature as a directly observable value. In this sense it is a *macroscopic* formulation, and the law is expressed in Clausius's formulation in the form of a dynamical law.

We need to change our approach to the thermal state if we turn to the microscopic or atomic structure of matter. The degree of a body's heat is determined by the energy of the motion of molecules according to the mechanical theory of heat. Molecules have different velocities and the energy is not evenly distributed between them. Both velocity and energy are distributed between the molecules according to a certain law.

Every given velocity distribution between molecules corresponds to a given thermal state of a body. In other words, each microstate determined by the distribution of molecules corresponds to a macrostate determined by temperature.

However, the same macrostate determined by temperature corresponds to *not one but several microstates*.

The same thermal state of a body can be realised by *different distributions* of velocity between molecules provided the mean energy remains the same. Indeed, it is irrelevant *which molecules* have this or that speed in order to determine the thermal state.

A given molecule's velocity is accidental. It is accidental in the sense of objective chance because it is the general character of the velocity distribution characterising the entire aggregate of molecules as a whole that is important in order to determine the thermal state, and not the individual distribution of velocities.

Similarly, it is the total distribution of numbers of heads and tails characteristic of the series as a whole being studied that is significant in the above example of the series of coin tosses and not the individual toss results.

Thereby, the same thermal state can be realised by a certain number of microstates.

This means that when we examine e.g. all possible microstates of a volume of gas paying attention to the velocity values of *each individual molecule*, we will get an extremely vast number of microstates. And earlier we saw that a whole series of microstates will realise the same thermal state.

The larger the number of microstates realising the same thermal microstate [sic],[19] the *more probable* this state is, in the same way as taking a white ball out of a box with 1,000 white and one black ball is more probable than pulling a black one.

Each thermal state has a certain probability from the *microscopic* [sic][20] *point of view.*

While observing the flow of thermal processes we are convinced that the difference between the macroscopic thermal states tends to level out, i.e. to reach the state of equilibrium. The process will not flow in reverse.

This is the essence of Clausius's formulation. Heat is always transferred from the warmer (i.e. at a higher thermal level) to the cooler body (at a lower thermal level).

Since we constantly observe only this direction of a process we conclude that *microprocesses realising the equilibrium state are more probable.*

If one gas volume is warmer than another then the thermal process will flow in such a way that temperatures will equalise, i.e. that the *most probable microstate* [sic][21] *is achieved, which is the state of thermal equilibrium.*

It is possible to artificially upset the thermal equilibrium by external intervention, e.g. by heating one of the bodies—in the same way that it is possible every time artificially to *choose* a black ball from the box. But if left alone, the thermal process will again arrive at the equilibrium in the same way that when we pull a ball from the box without choosing we get a distribution corresponding to the probability of pulling this or that ball.

Now it is clear why we consider thermal processes to be irreversible. They are irreversible because the transition from the less probable to the more probable state takes place during a thermal process. The probability of the reverse transition from the cold to the heated body is very low but not zero!

We observe no such a transition in nature because it is extremely unlikely but not at all because it is completely impossible.

If we put a pan on a burner the water will boil. We constantly observe this. But the water does not have to boil. It might also freeze, i.e. heat might be transferred from the

[19]TN: Presumably an error—should be "macrostate".

[20]TN: Again this seems to be an error—should be "*macroscopic*".

[21]TN: Presumably "*macrostate*".

water to the flame of the burner. It is not impossible but so very unlikely that the time required for a single occurrence of this state would be so long that by comparison the existence of our entire solar system is disappearingly short. Similarly, in order to pull a black ball from the box with 1,000,000 white and one black balls there would be an extremely long series of getting the balls. The black ball certainly does not have to be pulled out *last*. It can appear at any moment, even in the beginning, but the number of its appearances will be a million times less than that of a white one.

Generally speaking, an occurrence of any infinitely unlikely microscopic state is not impossible and can happen at any moment like the appearance of the black ball. However, were it to happen and be observed by us, this would in no means disrupt the main trend of heat processes, i.e. to proceed from the less probable to the more probable state.

This conception of thermal processes is expressed in the form of *statistical laws.*

Conversely, in case of dynamical laws we have a definite unequivocal behaviour of a process. A stone lifted above the ground must *necessarily* fall down due to gravity. This is necessarily so in every particular case.

If I put a pan on a stove *water can boil and can turn to ice.*

A general law for the flow of heat processes, i.e. the transition from less probable to more probable states says nothing about the flow of *single processes*. A statistical law is an expression relevant to the entire aggregate of processes.

The law of the dissipation of energy is a *statistical law*; thereby the contradiction between reversible and irreversible processes is resolved, i.e. every process is both reversible and irreversible. It is *irreversible* as a macroscopic process or an object of human practice because the probability of its reversibility is disappearingly small in comparison to the time of human and earthly practice. This is why we are not going to freeze water by putting it on a cooker and will not feed stoves with ice. And not because it is completely impossible but because the realisation of such processes is infinitely unlikely within the framework of our practice.

Yet every process is also reversible as a microscopic process because any infinitely unlikely microstate will necessarily be realised within cosmic time intervals.

The statistical interpretation of the law of energy dissipation removes the difficulty of the question of the beginning of the world in time. The onset of an equilibrium state is death only from the limited "earthly" point of view. From the cosmic point of view any infinitely unlikely formation or deviation from the thermal equilibrium is possible. Thermal death is the beginning of new life. The *world has no beginning and no end, neither in time nor in space.* This is the final conclusion of the statistical interpretation of the law of the dissipation of energy.

The *unity* of reversible and irreversible processes essentially means that irreversible (heat) processes are *based* on the mechanical motions of billions of molecules, and therefore, every thermal state is *fundamentally* reversible.

The *specificity* of irreversible processes means that the law of their flow is a statistical and not a dynamical law, and this allows us to speak about irreversible processes although the transition of heat from a cooler to a warmer body can occur in individual cases.

If we accept that chance is a subjective category we should thereby declare the specificity of irreversible processes also to be subjective. However, we learn from our daily experience that thermal processes are *practically* irreversible. And their irreversibility is not based on the statistical laws being incomplete knowledge but on the specific distinction between macroscopic heat processes and the mechanical motion of molecules. This qualitative distinction is confirmed by our practice and does not disappear if we know the detailed motion of every molecule and "eliminate chance as a consequence of our ignorance."

Chance and consequently, probability and statistical laws are as objective as is *quality*.

We can understand the essence of the law of energy dissipation only by having understood both the unity and the specificity of reversible and irreversible processes.

To summarise our reasoning.

We establish the existence of both reversible (mechanical) and irreversible (thermal) processes by direct observation of the visible world (macroworld). These processes seem to be divided by an impassable abyss.

The macroscopic formulation of the law of energy dissipation (according to Clausius) allows for no assumptions about the elementary structure of bodies. A body is regarded as an individual.

When we examine a gas from the atomistic point of view we establish that a gas is an aggregate of molecules and each of them moves according to laws of mechanics.

Mechanical motions are reversible. Thermal phenomena are irreversible. Mechanical phenomena underlie the thermal ones. How then can they lead to irreversible processes?

The answer is in the acceptance of the hypothesis of molecular chaos, whereby we interpret irreversible processes *statistically* and not dynamically (i.e. purely mechanically).

Thermal processes turn out to be both reversible and irreversible. Their specificity is established together with their unity.

A statistical interpretation became possible because the same thermal (macroscopic) state of a body is realised by a vast number of microstates.

Each one of the series of microstates realising a given thermal state is accidental in relation to the thermal state because this or that distribution of molecules forming the given thermal state within certain limits has no influence on it. Since we can regard known microstates as accidental we can form the concept of the probability of a given thermal state and thus arrive at the statistical interpretation of the law of energy dissipation.

If we were to regard chance as a subjective category, the law of energy dissipation (a statistical law) would acquire a subjective colouring. And then this law, unlike the law of energy conservation, would only be an expression of our ignorance of nature, and not an expression of an objective law.

However, by accepting the standpoint of dialectical materialism we view chance as an objective category, and thus arrive at the conclusion that natural science is founded on two equal laws, i.e. the fundamentally quantitative law of energy conservation and transformation and the law of energy dissipation. The latter fundamentally reflects the specific laws of the flow of energy, emphasizes the qualitative aspect of processes of this flow and thus compliments the law of energy transformation.

(To be finished in the next issue).[22]

References

Campbell, L., & Garnett, W. (1882). *The life of James Clerk Maxwell with the selection of his correspondence and occasional writings*. London: Macmillan.

Cournot, A. A. (1851). *Essai Sur Les Fondements de Nos Connaissances Et Sur Les Caractéres de la Critique Philosophique* (Vol. I). Paris: Hachette.

Engels, F. (1988). *Marx engels collected works* (Vol. 25). Moscow: Progress Publishers.

Gross, H. (1911). *Criminal psychology*. Boston, Massachusetts: Little Brown and Co.

Laplace, P. S. (1902). *A philosophical essay on probabilities*. London: Chapman and Hall.

Planck, M. (1960). *A survey of physical theory*. New York: Dover.

Poincaré, H. (2003). *Science and method*. New York: Dover Publications.

Quetelet, M. A. (1842). *A treatise on man and the development of his faculties*. Edinburgh: William and Robert Chambers.

Smoluchowski, M. (1918). Über den Begriff des Zufalls und den Ursprung der Wahrscheinlichkeitsgesetze in der Physik. (On the Concept of Chance and the Origin of the Laws of Probability in Physics). *Die Naturwissenschaften (The Science of Nature), 6*(17), 253–263.

[22]TN: See no. Chap. 6, p. 69, n. 1.

Chapter 9
(Selections from) The Main Ideas of the Theory of Relativity

Boris Hessen

Introduction

It is very difficult to present the theory of relativity in simple terms without the aid of mathematical expressions and without relying on readers' prior general knowledge of the most important areas of physics. A whole range of the main concepts of special relativity theory and particularly general relativity theory can be exactly expressed only in mathematical formulas.

Relativity theory came into being and developed as a physical theory within an elegant mathematical framework.

However, any fundamental physical theory regarding our principal views on nature always has a methodological basis.

Therefore, our task is not a detailed presentation of relativity theory but an identification of those methodological concepts that make up the foundation of its physical and mathematical constructions.

The present essay by no means embraces the entire contents of relativity theory but focuses primarily on the problem of space and time. No attention is given to the cosmological constructions of the theory as at present general relativity theory is undergoing a significant reconstruction.

Little attention is paid to the examination of the arguments against relativity because the author saw his main task in the positive presentation of the theory.

TN: Published by Moskovski Rabochii, Moscow, 1928.

B. Hessen (✉)
Abingdon, UK
e-mail: chris.talbot@protonmail.com

C. Talbot and O. Pattison (eds.), *Boris Hessen: Physics and Philosophy in the Soviet Union, 1927–1931*, History of Physics,
https://doi.org/10.1007/978-3-030-70045-4_9

The author would like to express his deep gratitude to Comrade A. M. Deborin and to Professor I. E. Tamm who have read this work in manuscript and made a lot of valuable suggestions.

B. Hessen
February 1928

Part of Chap. 2. Experimental Basis for Special Relativity (pp. 33–35) (On the Ether)

. . . Light is also an oscillatory motion that spreads wave-like in a special medium—the ether.

The ether may be compared to the air in our example with sound. It is either carried by the system which moves in it, e.g. the Earth and the objects on it (in this case naturally, we are unable to discover the movement of the Earth vis-à-vis the ether) or it is not carried by bodies (in this case we must discover the movement of the system vis-à-vis the ether).

A range of well grounded theoretical and *experimental* arguments lead us to the conclusion that the ether cannot be carried by moving bodies.[1]

In this case we would have been able to discover the movement of the system (e.g. the Earth) vis-à-vis the ether. In other words, Galileo's relativity principle will not be observed vis-à-vis optical (light) phenomena.[2] Theoretical calculations show that laws for light phenomena will be different for stationary and moving systems vis-à-vis the ether. Therefore, by making the required measurements we shall be able to determine whether the system is stationary or moving vis-à-vis the ether.

The determination of a system's (e.g. the Earth's) movement in relation to the ether is of great significance. Should this movement be discovered, we could with more authority conclude that we found *absolute* movement.

Indeed, the ether fills the entire space: the spaces between individual molecules that make up physical bodies, the interstellar space and the entire universe. According to classical physics, the ether is a carrier of all electromagnetic phenomena that play the major part in physics. "Empty space" means space that is filled with ether. We

---

[1]BH: Strictly speaking one can make three assumptions regarding the ether:
I. *The ether is fully carried by moving bodies*, so that the speed of the ether inside the moving bodies is equal to the speed of the body. This is H. Hertz's hypothesis. Should it be true, we would have been unable to discover the movement of bodies in the ether similarly to being unable to discover the movement of a railway carriage in experiments with the movement of sound when the carriage was closed, and the air moved with it. But a number of experimental results disagree with this hypothesis and call for its rejection.
II. The ether is *partly* carried by bodies. This is the Fresnel and Fizeau hypothesis that also has been unacceptable.
III. The ether is *not* carried by bodies *at all*. This is Lorentz's hypothesis. In this case we must discover the movement of bodies vis-à-vis the ether exactly like in our example with the moving railway carriage without front and back walls when we were able to discover its movement vis-à-vis the air. Throughout the text when we speak about the stationary ether we mean the ether of Lorentz that is not carried by bodies, i.e. the ether that is stationary.

[2]BH: Because the medium is not connected to the system.

do not know any matter more elementary than the ether.[3] The electrons—the tiniest known bricks of the universe interact with the ether and have been regarded by classical physics as "knots" in the ether. The ether is some sort of "primary matter". But if the ether continuously fills the entire infinite space and is not carried away by bodies, as has been proved both theoretically and experimentally, then there are no means available to us that can move the ether *as a whole*. The ether is at rest and is at an *absolute rest* because it *continuously fills the entire space* and there is no body that could be used as a reference system with regard to the ether: every material system (and physics like any natural science deals only with such systems) must be located in the ether.

Therefore, the ether is, as it were, a primary and absolute system of reference. And if one succeeds in discovering motion in relation to the ether, it could be called *absolute motion*. Thus, the experiments that allow us to discover the motion of a body (e.g. the Earth) relative to the ether would be of fundamental significance.

Such experiments similar to the above experiments on the movement of sound in the air in relation to a moving railway carriage, were performed by Michelson, an American physicist, in 1881 and produced a *negative result*.

Part of Chap. 4. Space, Time and Matter (pp. 64–69)

The criticism of Newton's concept of space and time from the standpoint of dialectical materialism resulted in the conclusion that space and time exist *not outside matter*, but *in matter*; matter is their true reality or their objective synthesis. It is not time in general or space in general that possess objective reality but concrete space and concrete time, i.e. moving matter.

However, apart from accepting the reality of the existence of abstract space and time *outside matter*, Newton's concept that has been fully incorporated into classical physics, is not satisfactory for another reason. It accepts the existence of space and time independently and *separately* from each other.

Classical physics believes that space and time should be examined as entities completely independent from each other. *Separateness of space from time* is not only the way in which we perceive the outside world but also the *form of existence of space and time.*

To oppose this view dialectical materialism suggests the concept of *the unity* of space and time. In our perception of the outside world and of moving matter *through our senses* we separate space and time. But in real moving matter, space and time are tied into one complex (synthesis). Matter exists not in two separate and independent forms—space and time forms—but in one space-time form. Space and time are not added mechanically but are inseparably tied into one synthesis in moving matter.

Space and time are synthesised in matter and are inseparably tied to each other. Every space exists in time because it really exists only as matter; every process in time is always also a process in space as it always is realised in the transformation of matter. We do not know other processes except those connected with matter. The

---

[3] TN: The previous two sentences are quoted by Josephson (1991, p. 243).

thinking process as the highest form of motion and the most complex one is a process in a human brain that is inseparably tied with other processes there and is another side of a *material* process.

We shall come back to the synthesis of space and time in moving matter and to the separation of space and time in our perception in Chap. 8 on the four-dimensional world. There we shall understand the essence of the synthesis and separation of space and time using concrete examples.

From the standpoint of dialectical materialism vis-à-vis space and time we shall inevitably arrive at the negation of the reality of absolute, empty and immovable (and therefore, separated from time!) space. By accepting the reality of space in *concrete matter* we accept the real existence of *relative but not absolute motion*. Indeed, any motion, even if it is only a mechanical one, is a motion of matter. Matter moves only in relation to matter. It cannot move in relation to space in general or absolute space, because the latter is an abstraction of thought and is not objective reality. It cannot move in relation to absolute space as the latter is not available to us in our *sensory perception* and fundamentally cannot be available in it directly or indirectly as it is a creation of our thought and an abstract conception and not objective reality.

Therefore, any real motion is relative motion.

We are unable and could not be able to sense abstract time as pure and empty [sic] duration. We know and observe only processes. We cannot compare the duration of a real process and "absolute empty duration" and define time intervals as intervals in relation to absolute time of pure duration because absolute time is the same sort of abstraction as absolute space. We know and observe only processes and therefore compare the duration of one process with the duration of another one. We always measure (and are able to measure) only *relative duration.*

The concept of relativity is rooted in the very concept of measurement. To measure means to relate one measured object to another one which has been accepted as a unit of measurement. In order to measure it is essential to have two objects with a mutual interrelation. Either of the two can always be accepted as a unit of measurement.

According to Engels cognition—is a *sensuous* measurement.[4] Cognition is a process of interaction between subject and object, therefore any cognition *has to have* an element of relativity. Real knowledge is always relative. However, this does not mean that we cannot make an approximation to absolute truth through relative knowledge.

In this aspect the views of dialectical materialism and of many proponents of the theory of relativity on the relationship between absolute and relative truth drastically differ. These proponents elevate the relativity of our knowledge to the level of a general principle of cognition and thus negate the possibility of making an indefinite approximation to absolute knowledge.

We shall come back to this in Chap. 7.

We see that the denial of the physical reality of absolute motion and the statement (rooted in Michelson's experiment) that we can cognise only relative motion—the statement that constitutes the main idea of the relativity principle—necessarily follows from a dialectical conception of space and time. But the very concept of space

---

[4]TN: *Dialectics of Nature*, Engels (1988, p. 516).

and time is, to an extent, a consequence of the main concept in dialectical materialism of the relation between the absolute and the relative. In physics Einstein's views on space, time and motion are a concretisation of the dialectical conception of space and time.

Resume of Chap. 4

Newton's concept of space and time is the basis of classical physics. Newton distinguishes absolute or true space and time from relative or ordinary space and time. Absolute space is unavailable for experiments and cannot be perceived by the senses. This is why in practical work we use relative space and time which we can observe in real matter and in processes. According to Newton, absolute space and time have objective reality. Absolute space exists as an empty receptacle for bodies. Absolute time exists as pure empty duration. Absolute space and time do not need matter for their existence. Matter may be brought into absolute space from outside and may be placed in absolute time.

From the standpoint of dialectical materialism Newton's absolute space and time are empty abstractions that do not correspond to any objective reality.

Moving matter is actual objective reality. Space and time are realised in it and only in it. They are inseparable from matter and matter is inseparable from them because space and time are forms of the existence of matter. We can study and compare only the motion of bodies in relation to each other and not the motion of objects in relation to absolute space, as Newton believed. Therefore, only relative motion actually exists. We cannot compare the duration of a process with the duration of absolute time, i.e. the duration that is pure, absolutely uniform and independent of any processes.

We can observe and compare only the duration of one process with another process.

Therefore, relative time is realised in a concrete process.

Furthermore, Newton's physics examines space and time as forms which are independent of each other and exist separately.

Dialectical materialism considers matter as the synthesis of space and time. Space and time are inseparably linked in matter and are separated there only in our sensual perception.

As far as physics is concerned the views of relativity theory on space and time generally coincide with the views of dialectical materialism on the relationship between space, time and matter.

Chapter 7. **Philosophical and Physical Relativism (pp. 104–114)**

We have arrived at the conclusion that the process flow (time measurement) and the size of space intervals (body length) depend on the condition of the system where the measurements are made.

Time and space acquire definition only if the system in relation to which the measurements are made, is indicated. This is the *relativity* of space and time, and this is why Einstein's theory is called the theory of relativity.

Our knowledge about bodies and processes significantly depends on the state of an observer. In this sense we speak of the relativity of the results of measurement of space and time.

Is this relativity of our knowledge fundamentally insurmountable, or is it possible to remove the influence of the state of an observer, and thus to make a step towards the absolute knowledge of nature? We face this question with regard to the implications of the special theory of relativity.

Let us formulate the problem of the relationship between absolute and relative knowledge in general form before moving on to the analysis of the above question.

Science has a task to study the external world that exists outside ourselves and independently of our cognition and perception.

The *form of perception* of objective reality depends without a doubt not only on the state of the perceived object of the outside world but also on the organisation and the state of the perceiving subject.

However, *the form and the method of existence of an object* depend neither on the state of a subject nor on its organisation.[5]

The task before our cognition is to know the outside world. What is the character of our knowledge of the outside world?

This question can have two different answers.

In order to know an object (of the outside world) there should be a subject which will cognise it. Cognition becomes possible only when both a cognising subject and a cognised object are available.

But no subject is required in order for an object to exist.

Therefore, cognition of the outside world is cognition of an independently existing object.

We move to the cognition of an object through a subject.

But our cognition does not become a purely subjective cognition because of this. That a subject is an essential condition for cognition does not entail at all that the *entire content of our cognition is subjective*. Our cognition as expressed in the forms of a cognising subject, has an objective content that is a true reflection of the properties of external reality. The outside world is cognisable as it is, and this cognition is disclosed through the interaction between a subject and an object.

In this interaction we proceed to the objective through the subjective; through the relative knowledge we approximate absolute knowledge.[6]

This is the solution of the question about the character of our cognition offered by dialectical materialism.

Relativism gives a substantially different answer to the same question.

Relativism accepts the same premise that cognition requires an available subject. However, relativism believes that cognition cannot leave the boundaries of the subject. We cannot ascend to absolute cognition and cannot approach it. At each

[5]TN: The last two paragraphs are quoted by Josephson (1991, p. 244). However he uses the term "facilities" rather than "organisation" of the subject.

[6]TN: The last sentence is quoted by Josephson (1991, p. 245).

stage we have only relative knowledge. While dialectical materialism sees the process of cognition of the outside world as a sort of a *spiral* movement or a constant approximation to the initial point or the object, relativism sees knowledge as circular movement: an object can be cognised via this movement from different sides or at different aspects. However, this cognition always stays relative and cannot be considered to be a consistent approximation to absolute knowledge.

Relativism does not consider a subject to be just a *condition* of cognition but also an insurmountable obstacle for complete cognition.

> Relativism as a basis of the theory of knowledge is not only recognition of the relativity of our knowledge, but also a denial of any objective measure or model existing independently of mankind to which our relative knowledge approximates (Lenin).[7]

According to dialectical materialism, a subject is both a condition and the only *method* for the consistent coverage and cognition of an object. The way towards absolute knowledge is through a subject.

We can only know in the process of cognition and interaction what sort of an object it is. Any other formulation of the question is meaningless.

An object must be examined not in an abstract form [of an object] but in a practical form or subjectively. This is the main premise of dialectical materialism.

The materialistic premise of acceptance of the objectivity of the world and its independent (from us) existence must not be understood in the sense that true materialism consists in the fundamental elimination of a subject. This is completely wrong.

This metaphysical absolutisation of the objective is the essential difference between dialectical materialism and any other materialism. This is why Marx pointed out in his first thesis on Feuerbach that the real (subjective) side of cognition was developed by idealism. But it was abstractly developed by it as well. This led to the absolutisation of the subject. According to Lenin, subjectivism differs from dialectics as follows:

> ". . . in (objective) dialectics the difference between the relative and the absolute is itself relative . . . for subjectivism and sophistry the relative is only relative and excludes the absolute".[8] "The materialist dialectics of Marx and Engels certainly does contain relativism, but is not reducible to relativism, that is, it recognises the relativity of all our knowledge, not in the sense of denying objective truth, but in the sense that the limits of approximation of our knowledge to this truth are historically conditional".[9]

Dialectics *includes* the aspect of the relativity of relativism *but is not limited to relativism.* It includes subjectivism but is not limited to subjectivism in its idealistic interpretation.

We can correctly approach the question of the relative character of our cognition *only after* a proper evaluation and understanding of the dialectical unity of subject and object.

---

[7]TN: English translation from *Materialism and Empirio-criticism*, Lenin (1977, p. 137).

[8]TN: English translation from *Philosophical Notebooks*, Lenin (1976, p. 358).

[9]TN: English translation from *Materialism and Empirio-criticism*, Lenin (1977, p. 137).

Cognition is a process of reciprocal relationship between subject and object, these two opposites. Therefore, a subject and an object acquire a true reality, a living existence only in the *process of reciprocal relationship*. An object does not oppose a subject as some indifferent thing-in-itself. Not at all. It reveals itself to a subject through *reciprocal relationship*. It is cognisable, and this cognition is an endless process of reciprocal relationship.

Engels says, "We cannot go beyond cognition of this reciprocal relationship as there is nothing cognisable beyond it."[10]

We accept this reciprocal relationship in our usual reasoning. We accept that we speak of nature and describe it in human categories but believe this to be a flaw and want to go beyond this reciprocal relationship and fundamentally to get rid of the subject. But this is a mistake because there is nothing fundamentally cognisable beyond the reciprocal relationship.

There is no need for any mystical super-subjects and super-observers in order to approach absolute knowledge. The true essence of an object is disclosed to a subject in the process of cognition.

> From the standpoint of modern materialism, i.e., Marxism, the limits of approximation of our knowledge to objective, absolute truth are historically conditional, but the existence of such truth is *unconditional*, and the fact that we are approaching nearer to it is also unconditional. The contours of the picture are historically conditional, but the fact that this picture depicts an objectively existing model is unconditional.[11]

As we have seen, the theory of relativity establishes the relativity of space and time. We have shown in Chap. 4 that as well as *relative* space and time given to us through our sensual perception, Newton's conception has introduced also *absolute* space and time not given to us through our perception; but relative time and space have to be defined in relation to the latter.

The theory of relativity removes abstract concepts of absolute space and time. Space and time are realised in matter. According to Newton, the motion of matter can occur not only in relation to matter (a body moves in relation to another body) but also in relation to absolute space; and this will be absolute motion. However, if space and time are realised only in matter, and absolute space and time are only abstractions of moving matter, then only motion of matter in relation to matter is possible, or as we said earlier, motion of one system of coordinates in relation to another. By removing the concept of absolute space and time we introduce relativity as a principle of the physical study of the world. Essentially, the physical relativity (relativism) of space and time means that the definition of a time process and a space distance significantly depends on the state of the observer (the system where the observations are performed).

---

[10]TN: Presumably this refers to Engels's statement, referring to Hegel, that reciprocal action is the true "causa finalis" (final cause) of all things. "We cannot go back further than to knowledge of this reciprocal action, for the very reason that there is nothing behind to know." *Dialectics of Nature*, Engels (1988, p. 512).

[11]TN: English translation from *Materialism and Empirio-criticism*, Lenin (1977, p. 136). Italics in original.

A natural question arises: is not the theory of relativity a physically concrete definition of relativism as a general philosophical concept?

Indeed, attempts have been made to connect the physical content of the theory of relativity with philosophical relativism and to reinforce philosophical relativism by arguments of the theory of relativity—a physical theory where the concept of the relativity of space and time plays a fundamental role.

But relativism as a philosophical concept is certainly not a methodological foundation of the theory of relativity.

The relative is a *stage* in cognition of the absolute. Therefore, criticism of Newton's concept of space and time is a substitution of the metaphysical absolute concept of abstract time and space by the concrete relative concept of space and time that is realised in matter.

If we stop here and maintain that all our further knowledge must have a relative character and will never step over the boundaries of this relativity, then certainly it will be easy to transform the theory of relativity into fundamental relativism.

However, relativity theory may be regarded as a stage in further knowledge of the outside world and of the moving matter. Absolute truth is composed of relative truths. Absolute cognition is the limit of relative knowledge [towards] which it indefinitely approximates.

When we measure time intervals separately from space intervals [then] the measurements' results significantly depend on the position of the observer. But separation of space and time is also an abstraction. Space and time are tied into one synthesis in moving matter.

The theory of relativity establishes both the inseparability of space and time from matter and the *inherence of space and time in each other.*

The *inherence of space and time in matter* leads us from criticism of the absolute of Newton's concept to establishing the *relativity* of space and time intervals.

The mutual *inherence of space and time and their synthesis* allow us to ascend to a higher stage of knowledge and to eliminate the impact of the state of an observer on the study of space-time objects, i.e. to make a further step on the way towards absolute knowledge of the outside world.

Thereby it is not at all necessary to consider the relativity of space-time intervals as a relativity which cannot be overcome. The point of the establishment of this relativity and its value for scientific research is that once we are done with the concept of absolute space and time (which are empty abstractions), this relativity will give us a correct *relative* cognition of the outside world which is a necessary link in the process of the movement towards absolute truth.

Now we will examine how space and time can be synthesised on the basis of relative data obtained from the process of the separate study of space and time, i.e. the problem of four-dimensional space.

This synthesis of space and time will allow us to overcome the relative nature of time and space intervals and to eliminate the influence of the system of coordinates on the measurement's results.

Resume of Chap. 7

The external world exists outside our consciousness and does not depend on it. No cognising subject is required for the existence of the external world (the object). But a cognising subject is essential for the *knowledge* of the external world. Philosophical relativism considers the availability of a subject, and therefore, a subjective element in cognition to be an insurmountable obstacle for the knowledge of the external world. Cognition of the outside world is certainly relative. A subject cannot step outside its limits and overcome the relativity of its knowledge. Any cognition is cognition from the "standpoint of a given subject". No cognition is (and cannot be) an unlimited approximation to the absolute knowledge of an object which has the same independent valid content for all subjects.

Dialectical materialism contrasts this standpoint with a fundamentally different relationship between relative and absolute knowledge. Any knowledge necessarily includes a subjective element. The process of cognition is a process of reciprocal relationship between subject and object. The object reveals itself in a fuller and fuller manner to the subject in this reciprocal relationship. At each given stage knowledge is relative but this relative knowledge is a step on the way to absolute knowledge. Cognition is a historical process where we ascend all the time from the lowest step of cognition to a higher one.[12]

A subjective moment in cognition is certainly not an obstacle for the achievement of generally valid knowledge. We know an object through a subject. From relative knowledge we approximate towards absolute knowledge.

Physical relativism is a recognition of the relativity of our concrete knowledge of nature. The essence of relativity theory is in the establishment of the relative character of time and space intervals. Their size significantly depends on the state of the observer. If we stop here and deny the possibility of overcoming this relativity but rather substantiate this statement by the arguments of philosophical relativism, then the theory of relativity transforms into fundamental relativism.

But this conclusion is certainly not a necessary consequence of relativity theory. On the contrary, in the concept of a four-dimensional world we see an attempt *to overcome* the relativity of space and time measurements and a next step on the way towards absolute knowledge of the outside world and of moving matter.

Chapter 10. Relativity Theory and Ether (pp. 158–169)

When we spoke earlier about the essence of Michelson's experiment and its significance for relativity theory, we made certain assumptions regarding the nature of light. We considered light as vibrations in an elastic medium which we called the ether. As the theory of electricity and magnetism developed further, it was shown that light is electromagnetic oscillations. There is no need here to delve further into

[12]TN: Josephson (1991, p. 245), summarises as follows: "All knowledge of necessity includes a subjective element. The process of cognition is the process of the interaction between the subject and the object. In this interaction the object is revealed to the subject more and more fully. At each given degree knowledge is relative, but relative knowledge is a stage on the path to absolute knowledge. The process of cognition is a historical process."

more detailed explanations regarding the electromagnetic nature of light. All we need now is to establish the fact of its electromagnetic nature. The ether is a carrier of electromagnetic phenomena, and since the latter form the foundation for all natural phenomena, the ether is a certain "primary matter" or a "fundamental carrier" which is the basis for all natural processes.

The mechanistic explanation of all natural phenomena was the main and principal task of theoretical physics in the nineteenth century. The mechanistic explanation of nature and the mechanistic worldview constituted the leading idea of physics from the second half of the seventeenth century onwards.

Here is how Huygens, one of the most prominent mathematicians and physicists of the seventeenth century formulated the principle of mechanistic explanation of nature:

> This is assuredly the mark of motion, at least in the true Philosophy, in which one conceives the causes of all natural effects in terms of mechanical motions. This, in my opinion, we must necessarily do, or else renounce all hopes of ever comprehending anything in Physics.[13]

This main premise setting out the essence of the mechanistic approach dominated in physics for almost two and a half centuries. It was thought that a phenomenon was finally explained and completely known when it was fully reduced to the motion of material particles. All properties of real matter had to be deduced from and reduced to the motion of the smallest elementary particles.

The mechanical explanation worked to a significant degree and led to extremely fruitful results in some areas of physics (kinetic theory of gases); but other areas, and primarily the area of electromagnetic phenomena, resisted the mechanistic explanation and could not be reduced to the motion and positioning of elementary particles, in spite of all the efforts by the best intellects.

We consider mechanical motion, the simple movement of a body as a whole in space, to be the most common and understandable. Terrestrial and celestial mechanics was the first discipline to develop. Therefore, it is natural to strive to reduce all natural phenomena to the mechanics of elementary particles.

Beyond doubt the mechanistic interpretation of nature is the simplest and the most obvious from the standpoint of "common sense". However, the things that common sense regards as the simplest turn out far from always to be true.

The mechanistic approach in physics started to fail already by the end of the nineteenth century when all attempts at the mechanistic interpretation of electromagnetic phenomena were unsuccessful. The relativity principle means the fundamental denial of the mechanistic approach and its supersession by significantly different methodological premises.

The denial of the mechanistic approach should certainly not be understood as a fundamental denial of atomic science in its entirety and of the kinetic theory of matter etc. Relativity theory certainly does not deny any of these achievements in physics.

---

[13]TN: Huygens (1969, p. 3).

However, *mechanistic atomic science*[14] as a general *methodological premise* for the entire edifice of physical science has indeed been questioned.

This is particularly clearly identified in the problem of the ether. If consistently applied, the mechanistic approach should deduce all physical phenomena, including the electromagnetic ones, from motions of the ether. All properties of material bodies also should be explained by the properties of the ether because all material bodies consist of atoms and atoms should be interpreted as formations in the ether.

However, here the difficulties begin. If the ether indeed exists as a medium similar to that known to us as liquid, gaseous and solid media (water, air and steel), and light is indeed oscillations in this medium, as are all electromagnetic phenomena, then the motion in relation to this medium must be discovered. But experiments failed to discover this motion.

How are we to reconcile this with the existence of the ether and is it possible to speak of its existence at all if firstly, the ether cannot be experimentally discovered and secondly, all attempts to reduce all phenomena to the motion of the ether—the "elementary matter", fail.

The properties of the ether as primary matter are completely unknown. In saying that light is electromagnetic vibrations of the ether we do not mean that by accepting the ether and by endowing it with very specific properties we then can obtain all laws of optics. The ether is just a carrier of electromagnetic phenomena but neither the properties of this carrier are known nor are they amenable to experimental study. Light from the Sun reaches the Earth in 8 min. If we ask where a ray of light is 4 min after it was radiated by a Sun atom, we should answer, "Somewhere between the Earth and the Sun". But if we consider light to be a process in a certain medium then global space must be filled with this medium. It then contains a ray of light before it reaches the Earth. It is in this sense that we speak of the ether as a carrier of electromagnetic phenomena.

The very fact that electromagnetic phenomena *require time for their distribution* means that they occur in a medium. At the same time this medium cannot be discovered by experiments and is not amenable to mechanistic interpretation.

The way out of these contradictions is twofold. We can fundamentally deny the ether as a useless hypothesis. Then there is a simple interpretation for Michelson's experiment: we cannot discover any motion in relation to the ether because there is no ether.

But this standpoint leads us to pure phenomenalism[15] and ultimately to idealism. Either we should speak in this case about an electromagnetic field[16] and fundamentally give up hope of resolving the problem of the *carrier* of electromagnetic fields or about a field (light oscillations) occurring literally in a vacuum.

---

[14]BH: We understand mechanistic atomic science as an approach which allows the reduction of all physical phenomena to the motion of discrete (separate) material particles.

[15]BH: The phenomenological approach is understood in physics as an approach that sets its task to describe the laws of phenomena but at the same time to fundamentally deny any general natural scientific worldview, including any mechanistic or any other rationale.

[16]BH: As a reminder, we call a part of space where any forces act a force field.

However, to speak about waves and oscillations which occur without a material carrier means to speak of motion without matter. And motion without something that moves is an empty abstraction of motion similar to empty space. Real motion is always connected with matter.[17]

In addition, to fundamentally deny the ether means to oppose the general theory of relativity according to which there is no empty space, i.e. space without any geometrical or physical properties. The key concept of the inseparability of space and matter contradicts the existence of empty space.

Einstein says,

> . . . according to the General Theory of Relativity space is endowed with physical qualities; in this sense, therefore, there exists an Aether. According to the General Theory of Relativity space without Aether is unthinkable; for in such space there not only would be no propagation of light, but also no possibility of existence for standards of space and time (measuring-rods and clocks), nor therefore any space-time intervals in the physical sense.[18]

Although Einstein introduced the concept of the ether, he endowed it with completely different properties compared to the ether of mechanistic physics.

We imagine bodies consisting of vast numbers of material particles. Even bodies we call continuous media, e.g. water, are also regarded as an aggregate of elementary particles. An elementary particle's motion is viewed as a simple mechanical change of place.[19] We try to reduce any *change in the state* of material bodies to laws of motion of elementary particles or to [laws of] their mechanical motion. A mechanical change of place is assumed to be something elementary and simple.

The concept of mechanical motion is applicable either to separate material bodies or to media that consist of elementary particles. The concept of motion as a mechanical change of place is inextricably linked with *discrete* or discontinuous bodies.

The concept of motion as a mechanical change of place is not applicable to an *absolutely continuous* body because it does not consist of particles, in which *each* of them can be *separately* studied in time.

But the concept of motion as *change of state* is applicable to such a body.

The task of mechanistic physics consisted in deducing the change in the state as a simple change of place, i.e. the mechanical motion of elementary particles.

But the question could be reversed: if the ether cannot be introduced as a molecular medium, then mechanical change of place cannot assume the role of the absolutely elementary type of motion. Although it seems simpler and more obvious, this is not a sufficient reason to make it the foundation for our entire approach. Conversely, we can accept change of state as a concept applicable to the ether in which the concept of a mechanical change of place is not. Then motion as a mechanical change of place would be *secondary*.

---

[17]TN: This paragraph is quoted by Josephson (1991, p. 243).

[18]BH: Albert Einstein, "Aether and the Theory of Relativity", Address delivered on May 5th, 1920, at the University of Leyden, Netherlands. TN: See Einstein (2010, p. 8).

[19]TN: The last phrase is quoted by Josephson (1991, p. 243).

This is the main difference between the ether of relativity theory and that of mechanistic physics. The ether of the principle of relativity does not consist of particles and does not have a molecular structure, therefore, the concept of motion as a mechanistic change of place is not applicable to it.

However, as it does not consist of particles it is impossible to discover a body's motion in relation to this *ether*.

Indeed, let us imagine a ship moving in ideally still water, water which it does not drag along. Let us further imagine an observer on board the ship who does not see any coast or bottom. In this case such an observer will be totally unable to discover the ship's motion in relation to the water although the ship undoubtedly moves in water—the medium where it moves. However, its movement cannot be discovered because for us the medium is continuous; and our method of observation does not allow us to discover the medium's atomic structure.

Therefore, the negative result of Michelson's experiment can be completely reconciled with the presence of the ether. However, motion in relation to it cannot be discovered because it does not consist of particles.

These are the views of the theory of relativity on the ether. The problem of the ether is one of the most complicated and difficult ones in physics. The general fundamental views on the ether expressed by the theory of relativity *by no means resolve the problem of the ether.* The latter is a key problem in mathematics and physics, i.e. the problem of discontinuity and continuity.

Mechanistic physics offered a solution in the sense of the priority of the discontinuity of the structure of matter—the science of the atom. Relativity theory suggests a concept of continuous matter. Further development of physics must arrive *at the synthesis of these two opposites* which are the main opposites in every area of theoretical physics.

The nature of the ether, i.e. the problem of the structure of matter is still far from [being] solved. Physics only approaches its solution. Demanding an a priori mechanistic explanation as the only possible and acceptable solution would be completely wrong and non-dialectical.[20] The *nature* of the ether is equally poorly explained by the theory of relativity and by mechanistic physics, but the ether plays an equally *fundamental* role in both.

Einstein says, "To deny the ether is ultimately to assume that empty space has no physical qualities whatever. The fundamental facts of mechanics do not harmonize with this view."[21]

The acceptance of empty space devoid of any physical properties means the objectification of Newtonian space as a mathematical abstraction. However, this abstraction has no analogy in the outside world. Real space existing outside our head, beyond us and independently of us, is realised only in matter.

The ether of the relativity theory is precisely the basic matter that *realises space.*

[20]TN: This sentence is quoted by Josephson (1991, p. 244).

[21]TN: See Einstein (2010, p. 6).

This ether is a carrier of electromagnetic fields; this is why there is no ground for accusing Einstein for thinking in terms of motion without matter. The ether's role as a carrier of fields is exactly the same in the theory of relativity as in mechanistic physics.

We have seen in the chapter on the General Theory of Relativity that the key concept of this theory is the inseparability of space and matter. The concept of space as an indifferent repository of matter contrasts with that of space whose structure is determined by matter.

According to Einstein's concept absolutely empty space does not exist and cannot exist because absolutely empty space cannot be a carrier of electromagnetic fields.

On the one hand, the Special Theory of Relativity does not consider the ether necessary but does not demand its fundamental denial; on the other hand, the General Theory of Relativity considers the ether necessary albeit with fundamentally different properties from the ether of mechanistic physics.

Resume of Chap. 10

The ether plays an important role in classical physics as a carrier of electromagnetic fields and a medium for the diffusion of light. However, despite this, the ether failed to be given a mechanistic interpretation; and despite all the efforts, a mechanical model of the ether was not built. In addition, experiments failed to discover motion relative to the ether which should have been discovered.

Two approaches appeared in physics thanks to the Special Theory of Relativity. The first adopts a purely formal standpoint and completely denies the existence of the ether. This standpoint leads to the necessity of motion existing without matter, and ultimately, to idealism.

The second approach shared by Einstein, believes that the ether must exist, but this ether in contrast to the ether of classical physics does not have an atomistic structure. Therefore, the concept of motion as mechanical change of place is inapplicable to this ether unlike the concept of motion as a general change of state [which it certainly is].

While the Special Theory of Relativity does not fundamentally oppose the acceptance of the ether, the General Theory of Relativity which denies the existence of "absolutely empty" space and endows space with physical properties and a structure, assumes that the ether must exist.

The problem of the ether is one of the most difficult problems in physics and was by no means finally solved by the theory of relativity.

The problem of the ether is essentially the problem of discontinuity and continuity, a cardinal problem in physics and mathematics.

The ether is fundamentally a carrier of force fields and matter realising space; and it plays the same role and is in the same way essential both in the theory of relativity and in classical physics.

Editor's Note—CT

Writing a non-mathematical introduction to relativity along with an exposition of related Marxist philosophy was surely a most ambitious project for Hessen. Unfortunately the whole book was too much for us to translate so we have settled for some of the most "philosophical" sections which were referred to by Josephson.[22]

The idea that space and time could be united in a dialectical synthesis is surely a powerful point for Marxist philosophy, as well as the conception that space and time must have their basis in moving matter. Newton's absolute "container" and universal time were clearly outmoded. These are points which Hessen was able to use in his argument.

There are though two problem areas. First the fact that special relativity includes the concept of an "observer", absent from Newton's approach. The observer is present even in the abstract form of a coordinate system. Doesn't scientific objectivity demand no dependence on the mind or on an individual's perception? In its developed form Marxism includes "human sensuous activity" and should therefore be a materialism that can include perception. Hessen understood this and we hope the translations here do justice to his efforts.

The second problem area is that of the "ether" or "aether". A materialist philosophy surely cannot accept an entity called "empty space" devoid of all matter and must put forward an ether or something similar. Yet the development of Einstein's special theory did seem to demand the absence of ether. This was not the case with the general theory which conceived space as being "curved", under the influence of matter, as Hessen points out. Still, apart from the very weak force of gravity, it seemed that the latest science denied any existence to the ether. This is why both Josephson[23] and Gorelik and Frenkel[24] criticise Hessen for his support for the existence of an ether, albeit a non-mechanical one. Non-materialist philosophies gained support with the widespread acceptance of special relativity. Many physicists supported some form of positivism, some, like British astronomer Arthur Eddington, even supported subjective idealism.

In the 90 years since Hessen was writing, physics has come round to support the existence of an "ether", or perhaps "ethers", though many physicists no doubt still wince at the use of the word. Let us quote from a fascinating non-mathematical popularisation of modern subatomic physics, *The Lightness of Being: Mass, Ether, and the Unification of Forces*, by Nobel-prize winner Frank Wilczek:

[22] Josephson (1991, pp. 243–245).

[23] Josephson (1991, p. 244).

[24] Gorelik and Frenkel (1994, pp. 50–53).

In the first part of the twentieth century, the upheavals of relativity and (especially) quantum theory shattered the foundations beneath classical physics. Existing theories of matter and light were reduced to rubble. That process of creative destruction made it possible to construct, over the second part of the twentieth century, a new and deeper theory of matter/light that removed the ancient separation. The new theory sees a world based on a multiplicity of space-filling ethers, a totality I call the Grid. The new world-model is extremely strange, but also extremely successful and accurate.[25]

And although Wilczek sometimes uses the term "empty space" it does seem to be anomalous given that he suggests it "is in reality a powerful medium whose activity molds the world."[26]

References

Einstein, A. (2010). *Sidelights on relativity*. New York: Dover.
Engels, F. (1988). *Marx Engels collected works* (Vol. 25). Moscow: Progress Publishers.
Gorelik, G. E., & Frenkel, V. Y. (1994). *Matvei Petrovich Bronstein and Soviet theoretical physics in the thirties*. Basel: Springer.
Huygens, C. (1969). *Treatise on light*. New York: Dover.
Josephson, P. R. (1991). *Physics and politics in revolutionary Russia*. Berkeley: University of California Press.
Lenin, V. (1976). *Collected works* (Vol. 38). Moscow: Progress Publishers.
Lenin, V. (1977). *Collected works* (Vol. 14). Moscow: Progress Publishers.
Wilczek, F. (2008). *The lightness of being: Mass, ether, and the unification of forces*. New York: Basic Books.

[25] Wilczek (2008, p. 10).
[26] Ibid., p. 73.

Chapter 10
Selected Material on the Work of Richard von Mises

Boris Hessen

The Statistical Method in Physics and the New Foundation of Probability Theory

1. A Collective and its Distributions

Formulation of the Problem

Exact natural science has been more and more conquered by the statistical method. However, in spite of its wide and fruitful application there is no sufficient clarity regarding its place among the number of other methodological tools and its methodological essence. And this is primarily reflected in the views on probability theory.

Probability theory is a mathematical tool for the study and expression of statistical laws. Therefore, it is only natural that our view of probability theory is determined by the evaluation of the character and significance of statistical laws in the general system of laws established by science. And conversely, the evaluation of probability theory leads to a certain view of statistical laws which it studies and expresses.

A very definite view of probability theory as an area of study where, due to the limitation of our knowledge, the full investigation of phenomena is impossible has prevailed since the time of its founders, Bernoulli and Laplace. We know something of the phenomena, but our knowledge is extremely incomplete. Laplace said, "Probability is relative, in part to this ignorance [of natural laws], in part to our knowledge".[1]

[1] BH: Laplace, *A Philosophical Essay on Probabilities*, 1908, p. 11. TN: Laplace (1902, p. 6).

TN: Translated from *Estestvoznanie i Marksizm (Natural Science and Marxism)*, 1, 1929. (Sections 1 and 2 and the Summary).

B. Hessen (✉)
Abingdon, UK
e-mail: chris.talbot@protonmail.com

C. Talbot and O. Pattison (eds.), *Boris Hessen: Physics and Philosophy in the Soviet Union, 1927–1931*, History of Physics,
https://doi.org/10.1007/978-3-030-70045-4_10

Probability theory is a result of our partial lack of knowledge. It does not *overcome* it but only partly makes up for it. This is why probability theory provides surrogate and not complete knowledge, and this makes it different from other mathematical disciplines.

According to Laplace probability theory is applied in the domain of chance. There is no fundamental or objective difference between the phenomena studied by mechanical methods and by those of probability theory. Each "curve described by a simple molecule of air or vapour is regulated in a manner just as certain as the planetary orbits; the only difference between them is that which comes from our ignorance."[2] We call chance a necessary phenomenon whose reasons we do not know. Chance is an unknown necessity and therefore, it has a purely *subjective* character.

This is why Laplace, by defining the area of application of probability theory as an area where chance is king, i.e. lack of knowledge, gives probability theory a subjective and transient character. The more knowledge we gain, the less is the area of application of probability theory.

Therefore, it is but a temporary crutch for our lack of knowledge and a surrogate for factual knowledge.

This subjective interpretation of probability theory traditionally has passed into modern science and thus made the definition of both probability theory's application and its tasks, and of the methodological value of its results, very difficult and obscure.

There is no doubt about the significance of R. von Mises's [contribution to the] precise limitation of the area of application of probability theory and his clear definition of its tasks and, as it were, his giving it completely "equal rights" with the other mathematical disciplines, and mainly his clear definition of the objects studied by probability theory. Von Mises's approach is completely opposite to that of Laplace. The subject of probability theory and the area of its application are determined by the objective properties of the object studied (collective) and not by our partial lack of knowledge.

The object of study by probability theory is defined as objectively, precisely and definitely as is the object of study by geometry. Its laws and conclusions regarding the object of study are as strict and precise as those of geometry.

The task of science is to study the objects and processes of the external world.

The study of phenomena makes it necessary to solve physical and mathematical problems and tasks. The character of these problems is determined by the specific structure of the studied objects. The methods of solution are most effective when they are best adapted to the characteristic structure of the studied object. Probability theory emerged and exists because the objects of a specific structure (collective) exist and their study results in the setting of quite definite tasks that require their study and solutions. Probability theory studies and finds solutions to the mathematical problems that are posed by the study of the area of phenomena where the collective plays the key role. Therefore, probability theory does not differ from other mathematical

---

[2]TN: Laplace (1902, p. 6).

disciplines in any way in the *origin* of problems, in their *processing* or in the *character of the results* that are to be used and realised in practice.

Von Mises aims at constructing probability theory in such a way that it would be a well founded and precise tool of the mathematical processing of problems that is as adjusted to the structure of the studied object as analysis or geometry.

A mathematical definition of probability is at the basis of the edifice of probability theory. It allows us to construct probability theory as a mathematical discipline. The definition of probability is the first principle[3] in Laplace's construction. This is a false definition because it contains an obvious *petitio principii*. The concept of probability is reduced to the concept of "equally possible cases". Equally possible cases are nothing other than equally probable cases. Notwithstanding the fact that this definition of probability is completely *inapplicable in practice* (we shall say more about this later) it is logically untenable.

According to Laplace, "cases equally possible, that is to say, to such as we may be equally undecided about in regard to their existence."[4] This definition follows from a subjective interpretation of the concept of probability and well illustrates the helplessness of the subjective concept in the rationalisation of probability theory.

If the objective interpretation of probability is used as a corner stone, as is done by von Mises, then the rationalisation and the construction of probability theory should not start with the definition of probability but with the *characteristics of those objects whose peculiar structure leads us to establishing the concept of probability*.

The definition of *probability* may and should be given only after specific features of the structure of the objects to which it refers are clearly established.

This approach to the problem allows us to precisely and objectively define probability and at the same time to unequivocally limit the area of application of this concept, and therefore, *the area of application* of theory of probability.

We shall look at the definition and examination of these specific objects (called *collectives* by von Mises) whose existence is at the basis of the construction of probability theory.

2. Statistical and Dynamical Laws. Additive and Non-additive Aggregates

As soon as the kinetic theory of gases began to develop, Maxwell, with his typical foresight, posed a question of the new type of laws introduced by this theory. In one of his lesser known philosophical essays he says, "But I think the most important effect of molecular science on our way of thinking will be that it forces on our attention the distinction between two kinds of knowledge, which we may call for convenience the Dynamical and Statistical."[5]

---

[3]BH: Laplace, *A Philosophical Essay on Probabilities*, 1908, p. 15. TN: Laplace (1902, p. 11).

[4]TN: Laplace (1902, p. 6).

[5]BH: Maxwell looks into the problem of necessity and chance applied to physics in a small article published by Campbell and Garnett – Maxwell's paper was given to a philosophical group at Cambridge (Club of Seniors). The article is titled "Does the progress of Physical Science tend to give any advantage to the opinion of Necessity (or Determinism) over that of the Contingency of Events and the Freedom of the Will?" TN: See Campbell and Garnett (1882, pp. 209–213). The original English is taken here from p. 210. See Chap. 4, p. 57, n. 45.

One of the specific and key features of the difference between dynamical and statistical laws is the fact that dynamical laws describe a *single or individual phenomenon* while statistical ones describe *an aggregate* of individuals or phenomena.

Although statistical laws cannot be used in the study of individual phenomenon because they simply make no sense with regard to an individual, the application of dynamical laws in the study of aggregates or collectives is not so simple. Every aggregate consists of individuals. Therefore, on first sight it would seem that dynamical laws apply well to the study of aggregates.

However, it is not so simple.

The same dual approach could be applied to statistical laws and to the concept of probability: an aggregate consists of a vast number of individuals. The behaviour of each individual is unequivocally determined by a dynamical law. We can study an aggregate as a sum of an immense number of dynamical laws. Such an approach is possible but extremely difficult. This is why we turn to statistical laws although they are second class knowledge in comparison to dynamical laws. However, the former do successfully overcome both the lack of our knowledge and partly, the extraordinary difficulties in studying aggregates arising from the immense number of individuals.

Typical arguments for the *subjective* approach to probability are easily recognised in this reasoning. In this sense statistical laws result from the limitation of our cognitive powers. The *objective* approach to statistical laws essentially means that the raison d'être of these laws is not in our limited knowledge but in the characteristic structure of aggregates or objects studied by them.

By using the statistical method in the study of an aggregate we examine it as a *whole*.

An aggregate, although consisting of separate elements, is not divided into separate elements in the process of our study. We study it as a whole or synthetically. Therefore, our statistical laws are applicable to the aggregate only as a *whole* and have no sense if applied to separate elements.

But what does it mean that statistical laws are valid for an aggregate as a whole and not applicable to its separate elements? In other words, what is the relationship between the dynamical laws that govern separate elements of the aggregate and statistical laws applicable only to the aggregate as a whole? Two fundamentally different approaches are possible here.

First: the laws that characterise an aggregate as a whole can be fully deduced from dynamical laws for the individuals. In this case, *in nuce* these laws are already expressed in each individual law. *Laws for the whole are a quantitative sum of laws for single phenomena.*

According to this approach an aggregate assumes a clearly additive character. It is divided without any remainder into the sum of its elements. There is a quantitative and not a qualitative difference between the whole and its parts. Laws for the whole are a sum of individual laws just like a kilogram is a sum of grams.

It is this concept of an aggregate as a purely additive formation that is the basis of the subjective interpretation of the statistical method. Indeed, if laws for the whole are simply a sum of laws for the parts then the sum of these laws is the only way to the precise knowledge of the laws for the whole. Since the law for the whole is a sum of laws for the parts then the law for the whole will be fully and completely known only through its cognition as a total sum of individual laws.

A dynamical law is a true and objective law. A statistical law is a roundabout way towards [obtaining] the knowledge of the law for the whole as a sum of dynamical laws, a temporary crutch for our ignorance, and although it allows us to move further and further forward, we move with the gait of a limping cripple and not of a healthy man.

A second approach is first of all based on the difference between the additive and non-additive properties of aggregates. The above approach is valid only with regard to specific properties, i.e. additive ones, of aggregates and not with regard to all their properties. Additive properties mean that an aggregate as a whole is a simple sum of parts, and the laws for the whole are a sum of the laws for the parts, just as the weight of a whole is the sum of weights of parts. There is nothing qualitatively new in the weight of the whole compared to the weights of its parts. The weight is equally easy to divide into components and put them together again. The only difference between the whole and the parts is the quantity.

However, apart from the additive properties (e.g. weight) an aggregate may have significantly different properties which are not included in each of its elements and belong only to the aggregate itself. When the aggregate disintegrates into its individual components these properties disappear in the same way as chemical properties of a compound disappear during its decomposition into components.

Non-additive properties are typical for an aggregate as a whole, and only for it. They are not virtually contained in its individual components. They are manifested only in the whole and *qualitatively* differ from the properties of individuals.

Statistical laws reflect and study precisely these non-additive properties and therefore, statistical laws by their essence clearly cannot refer to the single individuals which compose an aggregate. This is *not their flaw* but their characteristic *peculiarity* because they deal with the study of precisely those properties that are manifested only in the whole and *do not exist* in single elements.

By accepting that the presence of non-additive properties in aggregates is a peculiarity of their objective structure we thereby give statistical laws an objective character.

In this respect the relation between statistical and dynamical laws is the relation between the laws of the whole and of the part. Dynamical laws are individual laws. However, they are not sufficient for the study of the law of the whole because the whole has both additive and non-additive properties.

Statistical laws do not negate or contradict dynamical laws. They are necessary and valid in their area and dynamical laws in theirs.

M. Planck rightly notes that a dynamical law is a condition for the occurrence of a statistical law.[6] However, this does not mean that a statistical law is reduced without a remainder to a dynamical law. This is correct only if the whole is identical to the sum of its parts.

---

[6]TN: Planck (1960, pp. 56–68). See also Chap. 8, p. 101, n.5.

And if this is not so (and the existence of non-additive properties in an aggregate confirms the absence of this identical relationship), then a statistical law genetically arises from dynamical laws in the same way as a whole originates from a part; however, it is not composed of or decomposed into dynamical laws but is a qualitatively new formation only inherent in the whole and not in its parts. Therefore, a statistical law is not second-class knowledge compared to a dynamical law. It is a fully and equally valid method of cognition that results from the objective structure of the objects of study.

We have mentioned earlier that probability theory is the mathematical apparatus of the statistical method. The reasoning applied above to statistical laws can be applied to the concept of probability. Most importantly, probability characterises an *aggregate as a whole* and not its every single element.

Major mistakes in the interpretation of probability arise when one tries to apply the concept of probability to a single individual which is a part of a collective.

As von Mises correctly notes, this is the essence of Marbe's mistake[7]; the latter assumed that the longer only boys continue to be born, the higher is the probability of a boy being born again.

Probability characterises an aggregate as a whole. It does not refer to a single individual in the aggregate and cannot be deduced from it. Probability is a non-additive property of an aggregate.

However, the concept of probability cannot be applied to every aggregate that possesses non-additive properties.

Therefore, we now turn to the definition and description of properties of aggregates which are objects for the application of probability theory. According to von Mises' terminology we shall call these aggregates a *collective*, as opposed to other aggregates.[8]

Preface to On Causal and Statistical Laws in Physics[9]

Philosophical and methodological problems in physics featured rather prominently at the Fifth Congress of German physicists and mathematicians in Prague.

P. Frank's report "How modern physical theories contribute to the theory of cognition"[10] focused on the issues of epistemology. *Von Mises*'s report whose translation is published in this issue of our journal focused on the problem of a physical law.

The development of quantum mechanics resulted in a new approach to the interrelation between dynamical and statistical laws. The problem of a statistical method as a separate way of expressing physical laws became most urgent. The extraordinary

---

[7]TN: von Mises (1964, p. 184).

[8]TN: This concludes the first two sections of Hessen's article. The remaining sections are an exposition of von Mises' statistical work, closely following von Mises (1957).

[9]BH: A report given at the Fifth Congress of German physicists and mathematicians in Prague on 16 September, 1929. Naturwissenschaften 1930, transl. by E. L. Starokodamskaya. TN: Translated from *Uspekhi Fizicheskikh Nauk (Advances in Physical Sciences)*, X, No 4, 1930, pp. 437–439. The Preface is to a Russian translation of von Mises (1930).

[10]TN: Frank (1930).

development of statistical physics drew attention to the problems of the theory of probability. The classical explanation of probability theory based on the subjective concept of chance has clearly become unsatisfactory.

In his article below *von Mises* attempts to approach the relationship between statistical and dynamical laws using his concept of the probability theory based not on the subjective understanding of randomness but on the notion of a collective.

In his epistemological ideas *von Mises* generally joins *P. Frank* who adopts a Machist stance.

Von Mises's explanation of regularities in physics could not help but reflect his unsatisfactory epistemological views.

As to the cardinal problem of causality in physics, although *von Mises* does not adopt the extreme views defended by *Heisenberg* and *Dirac* who reject the elementary causality of physical phenomena, he still does not give a sufficiently clear explanation of this problem. However, he tries to demonstrate that the concept of statistical laws does not exclude the concept of causality.

Equally unacceptable are *von Mises*'s arguments on the relationship of physical and philosophical concepts which essentially echo *Frank*'s idea of "natural philosophy".

A healthy grain of protest against philosophical dogma and fantasies of the philosophy of science is completely devalued by the rough pragmatic approach to solving this problem. If we adopt *von Mises*'s view and agree that the sole task of philosophy is to adapt to the solutions of this problem given by physics, then it becomes unclear why we need a philosophical examination of the problem and why it is being discussed at a congress of physicists.

All this is because *von Mises* does not draw a distinction between philosophical dogmas and true philosophy.

The unsatisfactory and inconsistent character of *von Mises*'s philosophical conceptions result in his metaphysical explanation of the concept of limit and limiting case that play a large role in the uncertainty relation.

Von Mises does not pose the question of the relationship between an indefinite approximation to a given limit and a limit as initially given in every single act of observation. Therefore, no correct solution is found to the problem of divisibility and indivisibility, a very significant problem for the uncertainty relation.

Despite all these drawbacks the article by *von Mises* below is of great interest because it attempts to consider the problem of the relationship between statistical and dynamical laws using a new approach contrary to the theories that find the sole solution to the problem in the expulsion of causality from physics.

Editor's Note—CT

Von Mises's theory, like Smolukowski's, stresses the objective nature of probabilities. As Hessen explains in the first article above, he developed the concept of a "collective", a special kind of aggregate with "non-additive properties." Thus a probability relates to the collective and not to individual elements in it.

This is not the place to consider the history of statistics and the central role played by von Mises's approach, especially in the period that Hessen was writing.

We merely note that Andrey Kolmogorov, the Russian founder of the current, widely used, approach to probability and statistics, developed in the 1930s, based the application of probability on von Mises's ideas.[11] As von Plato explains, an unnecessarily dismissive attitude to von Mises seems to have developed in recent decades.

Hessen has to distinguish between von Mises's statistical methodology, as it could be applied in science, and his positivist philosophy. Just the fact that he wrote about von Mises was enough to foster attacks from Stalinist critics.

According to von Mises the theory of quantum mechanics showed that there was "indeterminism" from the macro-scale all the way down to the micro.

> Until recently, we thought that there existed two different kinds of observations of natural phenomena, observations of a statistical character [macro], whose exactness could not be improved beyond a certain limit, and observations on the molecular scale [micro] whose results were of a mathematically exact and deterministic character. We now recognize that no such distinction exists in nature.[12]

Even allowing for the Heisenberg principle, according to von Mises all measurements could now be seen to have a purely statistical character. Von Mises does not distinguish between the macro and the micro level.

Von Mises's philosophy had led, as Hessen writes above,[13] to a "metaphysical explanation of the concept of limit" which in turn supported his purely statistical approach to quantum theory. For Hessen understanding the concept of limit required the realisation that there is a relationship beween the "indeterminate approximation to a limit" and the actual limit that was "initially given." This is not very clear, let us attempt to explain.

It seems that Hessen is pointing to the fact that at the quantum level, a quantitative process could lead to a limit where there is a qualitative change in outcome. This outcome could include discrete or indivisible "jumps", for example in the energy levels of an atom which emits distinct characteristic spectral lines. Such jumps do not happen at the macro level.

Engels explains this "problem of divisibility and indivisibility" with the following example:

> If we imagine any non-living body cut up into smaller and smaller portions, at first no qualitative change occurs. But this has a limit: if we succeed, as by evaporation, in obtaining the separate molecules in the free state, then it is true that we can usually divide these still further, yet only with a complete change of quality. The molecule is decomposed into its separate atoms, which have quite different properties from those of the molecule.[14]

---

[11] von Plato (1994), Chap. 6, especially p. 179.

[12] von Mises (1957, p. 217).

[13] See p. 137.

[14] *Dialectics of Nature*, Engels (1988, p. 358).

References

Campbell, L., & Garnett, W. (1882). *The life of James Clerk Maxwell with the selection of his correspondence and occasional writings*. London: Macmillan.
Engels, F. (1988). *Marx Engels collected works* (Vol. 25). Moscow: Progress Publishers.
Frank, P. (1930). Was bedeuten die gegenwärtigen physikalischen theorien für die allgemeine erkenntnislehre? *Erkenntnis, 1*, 126–157.
Laplace, P. S. (1902). *A philosophical essay on probabilities* (F. W.Truscott & F. L.Emory, Trans.). London: Chapman and Hall.
Planck, M. (1960). *A survey of physical theory*. New York: Dover.
von Mises, R. (1930). Über Kausale und Statistiche Gezetzmäßigkeit in der Physik (On Causal and Statistical Laws in Physics). *Die Naturwissenschaften (The Science of Nature), 18*(7), 145–153.
von Mises, R. (1957). *Probability, statistics and truth* (p. 1928). Dover, New York: Second revised English edition. Original German edition published by Springer.
von Mises, R. (1964). *Mathematical theory of probability and statistics*. New York: Academic Press.
von Plato, J. (1994). *Creating modern probability, its mathematics, physics and philosophy in historical perspective*. Cambridge University Press.

Chapter 11
On the Question of the Causality Problem in Quantum Mechanics. Preface to the 1931 Russian Translation of Arthur Haas, Materiewellen und Quantenmechanik

Boris Hessen

To quote *Lenin*, the crisis in physics was that "matter has disappeared". First the electron theory and later the theory of relativity demanded a fundamental transformation of the concept of mass, which did not agree with the concepts in classical physics.[1]

Physicists faced a dilemma: either to reconstruct the old, habitual concepts of classical physics in accordance with new facts or to declare a new concept of matter with this "disappearance of matter".

Materialism in the sense of recognising matter as an objective reality essentially became a focus of the polemical struggle at the time. A strict distinction between the philosophical concept of matter and matter as a natural scientific category, as defined by *Lenin*, provided a way out of the crisis. It was not that matter had disappeared but the limit of our current knowledge of matter had disappeared (*Lenin*). Our knowledge penetrated deep into the atom. The indestructible and unchanging atom of classical physics ceased to exist. The metaphysical concept of unchanging elements was destroyed. The idea of development entered the world of atoms. Not only was the atom as complex as the solar system, but likewise, it had its own history.

[1] TN: *Materialism and Empirio-criticism* in Lenin (1977). Chapter 5, Sect. 2 is entitled "Matter Has Disappeared". That the limit of our current knowledge of matter had disappeared rather than matter itself is on p. 260.

TN: This was the 2nd Russian edition, translated by P. S. Tartakovsky from the 3rd German edition. It was published by the State Scientific and Technical Publishers, Moscow and Leningrad, 1931. The original German book was also translated into English in 1928 as *Wave Mechanics and the New Quantum Theory*, Haas (1928).

B. Hessen (✉)
Abingdon, UK
e-mail: chris.talbot@protonmail.com

C. Talbot and O. Pattison (eds.), *Boris Hessen: Physics and Philosophy in the Soviet Union, 1927–1931*, History of Physics,
https://doi.org/10.1007/978-3-030-70045-4_11

A dialectical transformation of the concepts of classical physics was based on new discoveries; thus, the idea of development entered the doctrine of the structure of matter.

Therefore, although the struggle was essentially focused on materialism it helped the dialectical philosophy penetrate physics. The idea of development is one of the main premises of dialectical philosophy. However, the doctrine of unity of opposites is its core and essence. "In brief, dialectics can be defined as the doctrine of the unity of opposites. This embodies the essence of dialectics" (*Lenin*).[2]

A characteristic feature of the modern crisis in physics is that the transformation of concepts leads directly to the law of unity of opposites. All development of modern physics is marked by two pairs of opposite concepts: discontinuity versus continuity (using the terminology of modern physicists, duality of waves and particles) and statistical versus dynamical laws.

The crisis in classical physics which Lenin wrote about left intact the main cornerstone of classical physics, i.e. the idea of continuity. The idea of discreteness was not in itself alien to classical physics: atomic theories and the kinetic theory of matter belonged to an area of physics dominated by the idea of discontinuity. Despite this, continuity *prevailed*. *Langevin* in his remarkable article "The Physics of Discontinuity" rightly noted that a major part of the laws of atomic physics could not be expressed in the language of differential and integral calculus designed to convey the concept of continuity.

> However, when studying systems composed of a vast number of individual phenomena, values determined by our measuring methods usually concern so many elements simultaneously in the form of a sum or an average value, that we may without appreciable error regard them as continuous.[3]

Thus, discontinuity seemed to be subject to continuity. Yet attempts were made to process it using the mathematical apparatus designed to interpret and develop continuity.

When dealing with the two contradictory concepts of continuity and discontinuity, classical physics kept the primacy behind continuity; similarly, of the two contradictory laws (the dynamical and statistical) classical physics awarded primacy to dynamical laws. From the time of Newton dynamical laws were considered the main and the ideal type of physical law.

---

[2]TN: The English translation given in Lenin, V.I., *Philosophical Notebooks* in Lenin (1976, p. 222), is "In brief, dialectics can be defined as the doctrine of the unity of opposites. This embodies the essence of dialectics, but it requires explanations and development." Hessen omits the last phrase.

[3]TN: "La physique due discontinu" in Langevin (1923, p. 190). French original (the first sentence is paraphrased by Hessen):

"Il est probable même que la plupart de ces lois ne pourront pas s'exprimer dans le langage du calcul différentiel et intégral, créé pour traduire analytiquement la notion de continuité.

Cet admirable instrument ne convient qu'à l'étude des systèmes accessible à nos sens et qui sont en général composés d'un nombre énorme d'éléments. Les grandeurs qu'atteignent nos moyens de mesure intéressent d'ordinaire tant d'éléments à la fois par somme ou par moyenne de grandeurs individuelles, que nous pouvons, sans erreur sensible, les traiter comme continues."

Statistical laws were [considered to be] second-rate knowledge. They were regarded as a stage in the study of phenomena. According to classical physics, true knowledge was achieved only when the phenomena fitted within the framework of dynamical laws.

Quantum theory introduced the idea of discontinuity into energy, one of the most universal concepts of physics, and placed discontinuity *on the same* footing with continuity. But only as quantum theory developed further the contradiction between discontinuity and continuity was revealed to its full extent.

The question of their interrelationship became even more important.

The contradiction between continuity and discontinuity in modern quantum physics is incomparably deeper than in classical physics. The apparatus for analysing the infinitesimal allowed an artificial interpretation of discontinuity to be made but did not provide positive results. Differential and integral calculus refused to work in the new quantum physics. Quantum theory required a new mathematical apparatus to be developed for the expression of discontinuity in the same way as the apparatus of classical analysis was developed for an adequate interpretation of continuity.

The new quantum physics made it more and more difficult to see discontinuity as subordinate to continuity. Slowly the idea became acceptable that the solution to the problem of discontinuity and continuity lies neither in the assertion of primacy of one over the other, nor in bringing one of these contradictory concepts alongside the other, but in their *synthesis*.

> If quantum mechanics has any peculiarity, it is that it does not decide between two modes of presentation (corpuscles and waves) ...but after the seeming victory of one, reinstates the other and combines both in higher unity—*M. Born*.[4]

The very course of development of physics acutely necessitated a solution to the problem in the spirit of dialectics.

Of course, the synthesis of discontinuity and continuity, of waves and particles, as outlined in modern wave theories of matter provided a solution to the problem, which was far from final. However, the *approach* to the solution was very different. Both waves and particles were sufficiently well known in classical physics but, according to *Jordan*, the problem of the interrelationship between electrons and the ether "already differed from the one at the time of *Lorentz*'s theory, when electrons were swimming in ether, to a certain extent like foreign bodies."[5]

The contradiction ("duality" according to the terminology of modern physicists) between discontinuity and continuity is closely associated with the contradiction between dynamical and statistical laws.

Up until now physicists hoped to reduce statistical laws to dynamical ones and to express the content of the new quantum mechanics using differential equations.

---

[4]BH: M. Born, Uber den Sinn der Physikalischen Theorien. Die Naturwissenschaften. No. 6, 1929. TN: Born (1929, p. 118). The English translation given here is in Born (1968, p. 35).

[5]BH: Jordan, Charakter der Quantenphysik. Die Naturwissenschaften No. 41, 1928. TN: Jordan (1928, p. 771). The German original is:

"...heute schon ein anderes Gesicht gewonnen hat, als in der Zeit der LORENTZ-schen Theorie, wo die Elektronen gewissermaßen als Fremdkörper im Äther schwammen".

When it was discovered that *Schrödinger*'s differential equations do not express dynamical laws and only allow for a statistical interpretation, the problem of the relationship between dynamical and statistical laws entered a new phase.

While classical physics considered the dynamical law to be the only type of law and tried to establish its primacy, modern quantum physics fell into another extreme and put forward the statistical law as the only type of law.

This "change of roles" was associated with the changing views of modern physicists on the significance and place of determinism in physics. By elevating the supremacy of statistical laws modern physicists started to reject the causality principle in its classical form.

"The principle of indeterminacy" (according to *Eddington*) in modern quantum physics meant that the causation of phenomena observed in the macro-world was only a statistical average of the total lack of causality of elementary phenomena. "No causal mechanism was expected to back statistical laws."[6]

Compared with classical physics, the question was posed upside down: if, previously, a statistical law was based on a dynamical law, now the foundation of a causal law was expected to be in the statistical law, "based on acausal elementary phenomena".[7]

Schrödinger expressed this point of view best in his article "What is a Law of Nature?"[8]

> Whence arises the widespread belief that the behavior of molecules is determined by absolute causality, whence the conviction that the contrary is *unthinkable*? Simply from the *custom*, inherited through thousands of years, of *thinking causally*, which makes the idea of undetermined events, of absolute, primary acausality, seem complete nonsense, a *logical* absurdity.
>
> But from what source was this habit of causal thinking derived? Why, from observing for hundreds and thousands of years precisely *those regularities* in the natural course of events, which, in the light of our present knowledge, are most certainly *not governed by causality*; or at least not so governed essentially, since we now know them to be *statistically* regulated phenomena.
>
> Therewith this traditional habit of thinking loses its rational foundation. In practice, of course, the habit may safely be retained, since it predicts the outcomes satisfactorily. But to allow this habit to *force* upon us the postulate that, behind the observed statistical regularities, there must be causal laws, would quite obviously involve a logically vicious circle. [Italics in English original]

---

[6]TN: Hessen presumably refers to Eddington (1928). This was an update of the Gifford Lecture Eddington gave the previous year to take into account Werner Heisenberg's uncertainty principle. Eddington writes of the "Principle of Indeterminacy" (p. 220 and p. 294). There is no exact reference to the sentence in quotes, but Eddington states that "the greatest triumphs of physical prediction have been furnished by admittedly statistical laws which do not rest on a basis of causality" (p. 298).

[7]TN: This exact phrase does not appear in Eddington (1928) but see previous footnote.

[8]BH: E. Schrödinger. "Was ist Naturgesetz", Die Naturwissenshaften, No. 1, 1929. TN: "What is a Law of Nature?" Schrödinger's Inaugural Address at the University of Zurich, December 9th, 1922. This address was not printed on the occasion of its delivery. The English text here follows the original manuscript from which the address was read: Schrödinger (1935, p. 115).

In chapter XI *Haas* speaks of the problem of "the causal and the statistical views in atomic physics"; here he unconditionally supports *Schrödinger*'s point of view, shared by most modern physicists.

> Such a (statistical) interpretation reconciles the idea that the direction of motion of an *individual* light quantum is *indeterminate* and *accidental* with the strict regularity shown by observable optical phenomena such as interference and diffraction….
>
> Since now a precise description of atomic processes in the classical sense is impossible, the causal principle also naturally loses its significance for physics. For this principle, according to which the exact knowledge of the present renders possible an exact calculation of the future, is meaningless when an exact knowledge of the present is unattainable. Causality must, therefore, according to the theories of *quantum mechanics*, be discarded indealing with the elementary processes of physics, and can only be accepted in the case of the probabilities to be ascribed on statistical grounds to these individual processes.[9]

Is this point of view correct? Do modern achievements in quantum mechanics really "cancel determinism"?

To understand this, let us have a look at the history of the concept of causality in physics, and at the meaning attached to it by classical physics.

Modern natural science owes its independence to its freedom from teleology. It accepts only the causal examination of nature.[10]

One of the battle slogans of the Renaissance was: "true knowledge is knowledge by causes" (vere scire per causas scire).[11]

Bacon emphasized that the teleological view is the most dangerous of the idola. The true relations of things are found in mechanical causation. "Nature knows only mechanical causation, to the investigation of which all our efforts should be directed."[12]

A mechanical conception of the universe necessarily leads to a mechanical conception of causation. *Descartes* laid down the principle of causation (ex nihilo nihil fit) as an "eternal truth."[13]

Mechanical determinism came to be generally accepted on English soil, although it was often interwoven with religious dogma (for instance, the "Christian necessarian" sect, to which *Priestley* belonged). This peculiar combination—so characteristic of English thinkers—is also found in *Newton*.

The universal acceptance of the principle of mechanical causation as the sole and basic principle for the scientific investigation of nature was brought about by the mighty development of mechanics. *Newton*'s Principia is a grandiose application of

---

[9]BH: See p. 141 below. TN: Hessen is referring to the Russian translation. English translation taken from Haas (1928, p. 83 and p. 85). Hessen's italics.

[10]This paragraph and the following six are repeated in "The Social and Economic Roots of Newton's *Principia*", Freudenthal and McLaughlin (2009, p. 67). See Chap. 4, p. 49, n. 17.

[11]TN: English original from Francis Bacon, *Novum Organum*, Book II, Aphorism II, Bacon (2000, p. 102). (Freudenthal and McLaughlin 2009, p. 67, n. 54.) See Chap. 4, p. 49, n. 18.

[12]TN: This quote was not located by Freudenthal and McLaughlin (2009, p. 67, n. 55). See Chap. 4, p. 49, n. 19.

[13]TN: Descartes (1983), Part I, Article 49, p. 22. See Chap. 4, p. 49, n. 20.

this principle to our planetary system. "The old teleology has gone to the Devil,"[14] but so far only in the realm of inorganic nature, of terrestrial and celestial mechanics.

The basic idea of the *Principia* consists in the conception of the motion of the planets as a result of the compounding of two forces: one directed towards the sun, and the other that of the original impulse.[15] *Newton* left this original impulse to God but "forbade Him further interference in His solar system."[16]

This unique "division of labour" in the government of the universe between God and causation was characteristic of the way in which the English thinkers interwove religious dogma with the materialistic principles of mechanical causation, as pointed out by *Plekhanov*.[17]

The acceptance of the modality of motion, and the rejection of moving matter as *causa sui* was inevitably bound to bring *Newton* to the conception of the original impulse.[18] From this perspective, the conception of divinity in *Newton*'s system is by no means incidental but is organically connected with his views on matter and motion, as well as with his views on space, in the development of which he was greatly influenced by *H. More.*[19]

It is at this point that the entire weakness of Newton's general philosophical conception of the universe becomes apparent. The principle of pure mechanical causation leads to the concept of the divine impulse. The *bad infinity* of the universal chain of mechanical determinism ends in the original impulse, thus opening the door to previously rejected teleology.[20]

However, upon having created the world and given an initial impulse to matter God handed the world over to the rule of mechanical causality. The world, where the law of gravitation applied, continued independently [of God].

In this respect *Newton*'s system was a truly complete system of physical causality, as *Einstein* stated in his *Newton*'s jubilee article.[21]

---

[14]TN: *Dialectics of Nature*, Engels (1988, p. 475) (Freudenthal and McLaughlin 2009, p. 67, n. 56). See Chap. 4, p. 49, n. 21.

[15]BH: Halley's Letter to Newton of 29 June 1686. TN: Turnbull (1960, p. 441). See Chap. 4, p. 50, n. 22.

[16]TN: Freudenthal and McLaughlin (2009, p. 67, n. 57), give the reference "Newton allowed Him the 'first impulse' but forbade Him further interference in his solar system" (*Dialectics of Nature*, Engels (1988, p. 480)). They point out that only the second clause of the sentence is in quotes in the Russian. See Chap. 4, p. 50, n. 23.

[17]BH: Plekhanov, *The Role of the Individual in History*, Vol. 8, p. 274 (In Russian). TN: English translation Plekhanov (1976, p. 284). Available via https://www.marxists.org/archive/plekhanov/1898/xx/individual.html, cited 13.03.20. See Chap. 4, p. 50, n. 25.

[18]BH: The initial impulse is the tangential component, of which *Engels* accused *Newton*. TN: See n. 14 above.

[19]TN: Henry More. English seventeenth century Cambridge Platonist philosopher.

[20]TN: See Chap. 4, p. 50, n. 28.

[21]TN: Einstein (1927)—an English translation appeared in the Smithsonian Annual Report for 1927 (see https://www.pbs.org/wgbh/nova/article/einstein-on-newton/).

Newton expressed the law of causality in mathematical terms and gave it the appearance still regarded by theoretical physics as the only possible formulation of the causality principle in physics.

The study of the micro-world and internal atomic processes, as well as the accumulation of new experimental data, resulted in the revision of the causality concept within the framework of physical research in the same way as the development of exact natural science resulted in the revolution in the understanding of space and time.

Newton's physics is the physics of the macro-world. Clearly the transition to studying micro-world phenomena required an innovative approach.

Laws in *Newton*'s physics were dynamical laws. Probability had no place in *Newton*'s mechanics. Each subsequent state was uniquely and necessarily defined by the previous one. Therefore, *Newton* formulated his laws as differential laws in the form of differential equations.

Thereby, *Newton* completed the preliminary causal research of nature.

When physics abandons a purely phenomenological view point and penetrates the depths of the micro-world, the old methodology becomes insufficient. This is essentially because dynamical laws become insufficient as a method of explanation of complex collective systems.

The question of statistical and dynamical laws was posed simultaneously with the development of kinetic theory of matter.

Maxwell was the first to pose this question in its general principle form in one of his lesser known essays:

> But I think the most important effect of molecular science on our way of thinking will be that it forces on our attention the distinction between two kinds of knowledge, which we may call for convenience the Dynamical and Statistical.[22]

The contradiction between statistical and dynamical laws is closely connected with the "duality" of the macro-world and the micro-world, as *Planck* justly noted.[23] It is for this reason the problem of statistical laws is at the forefront of modern physics.

---

[22]BH: Maxwell looks into the problem of necessity and chance applied to physics in a small article published by Campbell and Garnett—Maxwell's paper was given to a philosophical group at Cambridge (Club of Seniors). The article is titled "Does the progress of Physical Science tend to give any advantage to the opinion of Necessity (or Determinism) over that of the Contingency of Events and the Freedom of the Will?" TN: See Campbell and Garnett (1882, pp. 209–213). The original English is taken here from p. 210. See also Chap. 4, p. 57, n. 45.

[23]TN: Hessen is referring to Planck's speech, entitled "Dynamical and Statistical Laws", which he gave as rector of Berlin University, two days after the outbreak of World War One. Planck (1914). Cf. Mehra and Rechenberg (2000, p. 3), where the following translation into English is given: "…the dualism between dynamical and statistical [physical] laws seems to be most closely connected with the microscopic and macroscopic worlds, which we have to accept as an experimentally established fact. …hence we cannot but concede to the statistical laws the position which they deserve in the whole system of theoretical physics".

With regard to discontinuity and continuity classical physics fully and completely accepted the viewpoint of the primacy of continuity; likewise, in case of the relationship between statistical and dynamical laws, classical physics, as noted above, considered dynamical laws to be the true regularities.

A dynamical law is primarily based on the complete and unambiguous determination of a subsequent state by the previous one.

This determination is possible only if the system under consideration is either completely isolated from all external (vis-à-vis this system) influences, or if such influences could be neglected.

Therefore, when we state that the position and speed of a body at a given moment unequivocally and necessarily determine all its past and future behaviour, we thereby assume that there are no external perturbing forces. Accordingly, a dynamical law is an abstraction and an idealisation of real relations. The causal chain, determined by a dynamical law, is fully defined within itself. Each subsequent state depends only on the previous one.

In this case a dynamical law excludes any interaction, both between a causal chain of successive states, and between the entire chain and events outside it.

In order to predict a subsequent state, it is necessary to have full knowledge of the previous states, which lie entirely within the chain, and *nothing else*. Everything happening outside a given chain, expressed by the dynamical law under consideration, is as if it ceases to exist for this law. Or, in any case, the influence of external processes on a given dynamical law can be made arbitrarily small.

A statistical law is completely the opposite to a dynamical law.

A planet's position is *unambiguously and precisely* determined by its previous position and speed. When a coin is tossed for the tenth time, heads or tails show *totally irrespective* of the result of the ninth toss. While every subsequent state in a chain determined by a dynamical law depends on the previous one and nothing else, in a statistical series every subsequent state (the tenth toss of a coin) absolutely does not depend on the previous one. If we are given only an n-th toss, we would be unable to say anything about the result of the $n + 1$-th toss based *only on this knowledge.*

Therefore, the main premise of a statistical series is the supposition of no dependence of the subsequent state on the previous one.

Does this mean that a statistical series unlike a causal dynamical chain is acausal and non-deterministic? Not at all. Our belief that a previous toss of a coin does not determine the subsequent one means that the result of tossing is not determined by a serial number of the toss but by a set of causes which lie *outside* this order, i.e. in the initial coin orientation, initial impulse, ejector mechanism, air movement as the coin falls and the whole complex set of initial conditions of a single toss. Each single toss is fully determined by this set of conditions.

However, a statistical law does not consist in the study of a *single toss* but in the study of the entire aggregate of tosses. In this sense, one single toss is accidental in relation to the aggregate of tosses.

It is accidental not in the sense of being non-deterministic but in the sense that this or that order of heads or tails in a statistical series does not affect the general statistical law which governs the entire aggregate of tosses. In itself, every single toss is essential but accidental in relation to the entire aggregate of tosses.

In a dynamical law every subsequent state is necessarily determined by the previous one; there is no place for chance.

In a statistical law a single coin toss is accidental in relation to the whole statistical series, but the whole series of single coin tosses *necessarily identifies a statistical law*.

Chance is the antithesis not of causality, but of necessity. Therefore, *A. Haas* is incorrect in stating the following:

> Such an interpretation (explaining the intensity of light by the probability of the presence of a quantum of light at a given place – B.H.) reconciles the idea that the direction of motion of an *individual* light quantum is indeterminate with the strict regularity shown by observable optical phenomena such as interference and diffraction.[24]

A single quantum's direction of motion is accidental not in the sense that it is not determined but because a single quantum's behaviour is *not essential* for the entire aggregate of quanta; only the whole identifies a statistical law. However, the knowledge of this statistical law does not allow us to predict a single quantum's behaviour or a coin's behaviour in a single toss. This by no means proves the acausality of a single phenomenon. The fact that a statistical law does not predict a single phenomenon's behaviour in the aggregate of phenomena can neither be considered a defect nor can it prove the acausality of a single phenomenon.[25]

A dynamical law is the law of a single phenomenon, but it is unable to express the law of an aggregate. The fact that it does not and cannot express the law of an aggregate should not be considered its defect but a necessary consequence of the nature of the expressed law.

A statistical law is an expression of the specific nature of a collective as opposed to an individual or a single phenomenon. It expresses the behaviour of an aggregate as a whole and by its nature cannot refer to the individual phenomena that make up a collective.

Naturally, each aggregate arises genetically from single objects. Each statistical law is based on the aggregate of dynamical laws. Therefore, it would be incorrect to oppose statistical laws and determinism. A statistical law is a law *sui generis* that cannot be reduced to a simple sum of dynamical laws and in this sense, it is opposed to the dynamical laws. It is a special type of law that is inextricably linked to the peculiarities of the nature of a collective as opposed to a simple sum of objects.

Classical physics considered only dynamical laws to be the true regularities. Therefore, the causality concept in classical physics was identified with dynamical laws.

As noted above, the peculiarity of the modern problem is that unlike in classical physics, statistical laws are not subordinate to the dynamical ones, and do not stand alongside them but present the only method that is currently available in physics and that is able to express the internal atomic phenomena's laws.

---

[24]TN: English translation taken from Haas (1928, p. 83). Hessen's italics.

[25]TN: This paragraph is quoted by Josephson (1991, p. 268).

This results in the following chain of conclusions: causality might be expressed only in the form of dynamical laws; quantum mechanics' development is characterised by the fact that statistical laws become the governing type of laws of elementary phenomena; but contrary to dynamical laws statistical ones do not determine single phenomena; this is why dynamical laws must be abandoned for elementary phenomena of the micro-world and the causality principle should be abandoned accordingly.

Obviously, this conclusion is mainly based on the identification of the causality principle as the fundamental principle of natural research with its individual expression in the form of a dynamical law and with merging the concepts of chance and acausality.

These conclusions follow only from contrasting a dynamical law as an exclusive law, with a statistical one.

Both a pure statistical law that assumes complete independence of the subsequent member of a statistical series from the previous one and a dynamical law are an idealisation.

Neither pure statistical laws nor pure dynamical ones exist in nature because neither simple isolated systems nor independent phenomena exist in nature.

However, real phenomena require the study of significant regularities; the difference between dynamical and statistical laws is rooted in the real nature of the objects under consideration.

A planet's motion is a dynamical law because we neglect its interaction with its environment. We would observe random variations similar to the ones we observe for a tossed coin if we look at an actual trajectory of a real planet and not at a trajectory of a physical point in mechanics; the aggregate of these variations can be expressed by a statistical law.

A series of coin tosses is characterised by the fact that the interaction between the coin and the ejector and the environment is manifested in the series.

A collective differs from a simple sum of objects precisely by the fact that the single objects which comprise the collective, interact in a complex way. This interaction results in a statistical law. Therefore, it does not fit within the framework of a dynamical law that examines cause and effect as externally connected but not interactive phenomena.

Dynamical laws correspond to mechanical causality, i.e. an abstract determinism that not only excludes interaction but requires full equality of all causal chains because dynamical laws do not make a distinction between significant and insignificant laws.

But if all causal chains are equal, if "a particular pea-pod contains five peas and not four or six, that a particular dog's tail is five inches long and not a whit longer or shorter, that this year a particular clover flower was fertilised by a bee and another not, and indeed by precisely one particular bee and at a particular time, …that last night I was bitten by a flea at four o'clock in the morning, and not at three or five o'clock, and on the right shoulder and not on the left calf – these are all facts which have been produced by an irrevocable concatenation of cause and effect, by an unshatterable necessity of such a nature indeed that the gaseous sphere, from which the solar system was derived, was already so constituted that these events had to happen thus and not otherwise", then the study of nature becomes impossible. "With this kind of necessity, we likewise do not get away from the theological conception of nature. Whether with *Augustine* and *Calvin* we call it the eternal decree of God, or Kismet as the Turks do, or whether we call it necessity …"[26] None of these cases call for studying causality, and in none of these cases can we move forward.

As we have seen earlier, a statistical law is closely linked to the concept of chance, i.e. to the difference between significant and insignificant laws. Statistical laws for a molecular collective are based on the molecules' interaction. A single molecule's behaviour is accidental vis-à-vis the laws of a collective as a whole as both motion and position of a single molecule are inessential for the whole collective's behaviour and for the laws of the whole aggregate.

Statistical laws are richer in content than dynamical ones as they rise above the concept of abstract determinism and imply [the concepts of] randomness and interaction.

Therefore, if we reject the fatalistic concept of determinism on the one hand and accept chance as not simply a consequence of our ignorance but an objective category, then the opposition between dynamical and statistical laws is destroyed.[27] They do not exclude but imply each other. They are both necessary and valid. A statistical law is not a consequence of our lack of knowledge of processes, but an objectively necessary research method rooted in the peculiarities of the phenomena under investigation. *Engels*'s concept of chance and necessity is the key to solving the problem, not by rejecting causality but by a correct synthesis of chance and necessity, and therefore, of dynamical and statistical laws.

A final state is always a necessary result of elementary processes. But a single elementary process is accidental in relation to the entire process in an aggregate.

This concept of a statistical law is based on the synthesis of chance and necessity; it ascribes chance an objective value and allows us to understand a statistical law always using a causal approach.

Nevertheless, a mechanical concept of causality and of an abstract dynamical law should be considered insufficient.

It is important to note that a governing concept in modern physics is that of a statistical law as a law based on purely acausal and absolutely undetermined phenomena.

At the fifth Solvay conference *Dirac* and *Heisenberg* defended the idea of a lack of any determinism in elementary individual processes that constitute the basis of statistical laws. *Dirac* stated that nature "makes a free choice" in elementary phenomena.[28] This approach is essentially a fundamental denial of the possibility of cognition of the elementary phenomena.

---

[26]BH: *Dialectics of Nature*, Archive of Marx and Engels, v. 2, p. 193. TN: *Dialectics of Nature*, Engels (1988, pp. 480, 499).

[27]TN: This sentence together with the previous two paragraphs are quoted by Josephson (1991, p. 268).

[28]TN: Bacciagaluppi and Valentini (2009, p. 495). The adjective "free" is not in the original.

Lorentz made a particularly sharp comment while concluding the discussion on causality, determinism and probability. He said:

> There is then, it seems to me, a fundamental difference of opinion on the subject of the meaning of these choices made by nature. (*Lorentz* is referring to comments made by *Dirac* and *Heisenberg* – B.H.) To admit the possibility that nature makes a choice means, I think, that it is impossible for us to know in advance how phenomena will take place in the future. It is then indeterminism that you wish to erect as a principle. According to you there are events that we cannot predict, whereas until now we have always assumed the possibility of these predictions.[29]

Modern physicists draw similar conclusions rejecting determinism of elementary phenomena on the basis of another fundamental difficulty of quantum mechanics known as the *Heisenberg* uncertainty principle, as well as incorrectly opposing causality and statistical laws allegedly based on acausal phenomena.

The *Heisenberg* principle asserts that the position and momentum of an elementary particle are the quantities that fully determine the initial conditions required in order to express the behaviour of that particle under a dynamical law; these quantities cannot be determined independently with any required precision.

An increased precision in the determination of a particle's position is related to an increased lack of precision in the determination of the momentum and vice versa; the lack of precision is of the order of *Planck*'s quantum of action.

A dynamical law requires the knowledge of initial conditions fundamentally with any given precision. If, as in quantum physics, initial conditions can fundamentally be determined only with a precision up to h, then the dynamical law loses its absolute precision and uniqueness. Since most modern physicists believe that dynamical laws are associated with the causality principle, the fundamental inability to determine exact initial conditions for them would be equivalent to a denial of the causality principle.

In his inauguration speech on the occasion of his election to the Prussian Academy of Sciences on 4 July 1929 Schrödinger worded this difficulty as follows[30]:

---

[29]BH: *Electrons et Photons. Rapport et Discussion du Cinquième Conseil de Physique, de l'Institut International de Physique Solvay* (Paris: Gauthier-Villars, 1928). TN: English translation taken from Bacciagaluppi and Valentini (2009, pp. 497–498).

[30]BH: Schrödinger, E., Sitzungsberichte der Preussischen Akademie der Wissenschaften, 1929. TN: Schrödinger (1984, p. 304). German Original:

"Eine der brennendsten Fragen, die uns in deisem Zusammenhang heute beschäftigen, ist die, ob mit der klassischen Mechanik auch ihre Methode aufzugeben sei, der Grundsatz, daß feste Gesetze im Verein mit den zufälligen Anfangsbedingungen das Geschehen im Einzelfall eindeutig bestimmen.

Es ist die Frage nach der Zweckmäßigkeit des unverbrüchlichen Postulates der Kausalität.

Praktisch hatt man auf die Kausalität allerdings schon im Rahmen der klassisch-mechanischen Naturerklärung verzichten müssen.

...

Die Unbestimmtheit entspringt dabei nur aus der praktischen Unmöglichkeit, den Anfangszustand eines aus Billionen von Atomen zusammengesetzten Körpers genau festzustellen.

Heute dagegen werden Zweifel an der eindeutigen Bestimmtheit des Naturgeschehens in ganz anderem Sinne laut.

> The most burning issue for us today is whether we should abandon, along with classical mechanics, also its basis and method i.e. that immutable laws unambiguously determine the outcome in each individual case depending on arbitrary values of the initial conditions. It is a question of whether it is expedient to preserve the inviolability of the causality postulate.[31]
>
> While in classical mechanics the accurate determination of initial conditions was impossible in practice, there was no doubt that [initial] conditions for any number of objects could be determined in principle with any degree of accuracy. In new quantum mechanics the determination of initial conditions becomes not only practically difficult but fundamentally impossible. It is impossible both for a complex system and for a single atom or molecule.
>
> For a natural scientist a phenomena that is fundamentally unobservable in principle does not exist; so the meaning of this [impossibility of determination] could be explained as follows: the state of even an elementary object cannot be established in such a way that a given impact [upon the object] will make the object behave in a certain determinate way.

The *Heisenberg* uncertainty principle revealed with striking clarity the limitations of the mechanical causality principle and of the necessary introduction of the interaction principle.

In classical physics the fundamental ability to measure a state with any degree of accuracy tacitly implies that the influence of the measuring apparatus on the measured object fundamentally can be reduced to zero. This assumption in the physics of the macro-world is justified because masses and energies of macro-world objects are so large that the measuring apparatus's influence can be neglected.

However, with internal atomic phenomena the [values of] masses and energies are comparable in order of value with h; this makes it fundamentally impossible to neglect the influence of the measuring apparatus, e.g. the influence of a beam of light on a moving electron. The measurer and the measured are of the same order, and currently there is no available apparatus whose order of value would allow us to neglect its influence on the measured object.

Consequently, in the micro-world the measurer and the measured are always given to us in interaction. The uncertainty principle indicates that in this interaction it is impossible to separate the measured from the measurer.

This situation is clearly described by *Bohr* in his famous speech at the conference in Como published in a separate article.[32]

> Now the quantum postulate implies that any observation of atomic phenomena will involve an interaction with the agency of observation not to be neglected. Accordingly, an independent reality in the ordinary physical sense can neither be ascribed to the phenomena nor to the agencies of observation.

Die Schwierigkeit bei der Festellung des Anfangzustandes soll nicht nur eine praktische, sondern eine prinzipielle sein, sie soll nicht nur für ein kompliziertes Gebilde, sondern schon für das einzelne Atom oder Moleküle vorliegen.

Da das prinzipiell nicht Beobachtbare für den Naturforscher als Naturforcher nicht existiert, ist der Sinn dieser Meinung: schon der Zustimmte Einwirkung ein ganz bestimmtes Verhalten des Gebildes nach sich zieht".

[31]TN: Hessen puts the following two paragraphs in quotes also. They are translated here from the Russian and only follow the original German loosely.

[32]TN: Bohr's lecture on the present state of the quantum theory delivered on Sept. 16, 1927, at the Volta celebration in Como. Its contents were published in the next reference.

> ...On one hand, the definition of the state of a physical system, as ordinarily understood, claims the elimination of all external disturbances. But in that case, according to the quantum postulate, any observation will be impossible, and, above all, the concepts of space and time lose their immediate sense. On the other hand, if in order to make observation possible we permit certain interactions with suitable agencies of measurement, not belonging to the system, an unambiguous definition of the state of the system is naturally no longer possible, and there can be no question of causality in the ordinary sense of the word.[33]

Therefore, the difficulty is rooted in the fact that we can examine objects only in their interaction and currently have no means for an isolated observation.

The uncertainty principle points to an internal link between a particle [sic] and a momentum and signifies a basic contradiction between continuity and discontinuity.

In order to determine a point, i.e. a purely discrete quantity, we use a wave process, i.e. a continuous quantity.

Many physicists consider the *Heisenberg* uncertainty relation to be a methodological research principle, so-called "fundamental unobservability in principle".

As modern quantum mechanics believes it impossible to determine the position and momentum of an elementary particle with any degree of precision, then consequently no means are available in order to observe an exact initial position of the particle in a physical experiment. Physics must deal with fundamentally observable quantities. Since the inability of observation of an elementary particle is not a consequence of the imperfection of our apparatuses but is a fundamental impossibility then physics in principle should not deal with such concepts as the position and velocity of a particle.

But the determination of a particle's position and velocity are necessary conditions for the research of its behaviour in space and time; if such a determination is considered impossible then in accordance with the "fundamental unobservability in principle" we should abandon our research of particles in space and time and limit this research to statistical statements about final states.

This is why *Bohr* says that "space and time lose their direct meaning".

Obviously, a real object should be available for observation. The question whether *a given* physical quantity is observable or not cannot be resolved a priori but only in accordance with a definite physical theory.

According to *Newton*'s theory an absolute empty space and an absolute acceleration are fundamentally observable quantities; according to relativity theory these quantities are not observable. This does not mean that relativity theory simply rejects the very existence of space. While accepting the impossibility of the observation of absolute acceleration relativity theory, instead of ascribing an independent real existence to an empty space separately from an empty time [sic], acknowledges the reality and observability of the synthesis of time and space in matter.

"Fundamental unobservability in principle" is a certain consequence of the quantum postulate and the role played in it by *Planck*'s constant. As in classical atomic physics, if we accept that an atom is indivisible we cannot consider half an atom to be

[33]BH: *Die Naturwissenschaft*, 16, 245 1928. Russian translation in *Uspekhi Fizicheskikh Nauk*, v. 8, 3. TN: The English translation used here is Bohr (1928, p. 580).

a fundamentally observable quantity; thus, by accepting an atom of action (quantum of action) we inevitably arrive at the impossibility of determining the position and momentum with any required degree of precision. However, this does not mean that this physical theory should be considered absolute, nor does it mean that one of its principles should be declared a supreme principle of natural scientific research.

One of the peculiarities of modern experimental physics is the fact that in some areas we are very close to the limit of the measurements' precision currently available in physics; the limit results not only from the technical difficulties in research or in the manufacture of apparatuses but also from the fundamental conditions of measurement. As shown by *Zernike* and *Ising* in their remarkable research the electric current's measurement is limited by the material which must be used for the manufacture of the apparatus, given Brownian motion and fluctuation of electrons.[34]

In order to push further the boundaries of research new states of matter are needed.

Undoubtedly the uncertainty principle is related to the limitations of modern experimental physics. However, it would be wrong to absolutise the current state of physical theory and consider this principle [i.e. *Heisenberg*'s] to be a postulate for all physical theory while it is merely a consequence of certain premises related to the current stage in the development of physics. This was stressed by *Lorentz* in his above speech at the Solvay conference.

> Examples given by Heisenberg demonstrate that I have reached all the results that an experiment can give me.
>
> But I believe that the concept of probability should be regarded as a final result of theory and not an a priori axiom; however, I willingly admit that this uncertainty agrees with the experimental possibilities.[35]

If we consider the uncertainty relation not as an a priori principle of all physical research but the *limit of current knowledge of matter*, then instead of a general epistemological principle it becomes a certain physical principle that signifies one of the stages in the development of physics in the same way as what was once an eternal and indivisible atom was a step towards the modern complex and historically developing atom.

The development of physics over the last quarter of a century has clearly demonstrated that such a metaphysical absolutising of individual physical principles and theories is unacceptable.

Space and time as forms of existence, as well as causality shall always be the supreme principles of natural research because they are not just forms of our thinking but objective forms of real existence and interaction of the objects of nature.

---

[34]TN: Beller (1999, pp. 92–95) suggests that the work of Zernike and Ising, demonstrating the limits of accuracy in real measurements, "provided Heisenberg with the crucial clue [in his uncertainty paper]: he would pursue an intuitive interpretation of quantum mechanics through an analysis of the limits of measurement in his thought experiments (p. 95)".

[35]BH: Ibid. TN: English translation taken from Bacciagaluppi and Valentini (2009, pp. 478–9).

References

Bacciagaluppi, G., & Valentini, A. (2009). *Quantum Theory at the Crossroads, Reconsidering the 1927 Solvay Conference*. Cambridge: Cambridge University Press.

Bacon, F. (2000). *The new Organon* (L. Jardine & M. Silverthorne, Trans.). Cambridge: Cambridge University Press.

Beller, M. (1999). *Quantum dialogue: The making of a revolution*. Chicago: University of Chicago Press.

Bohr, N. (1928). The quantum postulate and the recent development of atomic theory. *Supplement to Nature*, 580–590.

Born, M. (1929). Über den Sinn der Physikalischen Theorien. *Die Naturwissenschaften [The Science of Nature]*, *17*(7), 109–118.

Born, M. (1968). *Physics in my generation*. Heidelberg: Springer.

Campbell, L., & Garnett, W. (1882). *The life of James Clerk Maxwell with the selection of his correspondence and occasional writings*. London: Macmillan.

Descartes, R. (1983). *Principles of Philosophy* (V. R. Miller & R. P. Miller, Trans.). Dordrecht: Reidel.

Eddington, A. (1928). *The nature of the physical world*. Cambridge: Cambridge University Press.

Einstein, A. (1927). Newton's Mechanik und ihr Einfluß auf die Gestaltung der theoretischen Physik. *Die Naturwissenschaften [The Science of Nature]*, *15*(12), 273–276.

Engels, F. (1988). *Marx Engels collected works* (Vol. 25). Moscow: Progress Publishers.

Freudenthal, G., & McLaughlin, P. (Eds.). (2009). *The social and economic roots of the scientific revolution: Texts by Boris Hessen and Henryk Grossmann* (Vol. 278). Boston studies in the philosophy of science. Berlin: Springer.

Haas, A. (1928). *Wave mechanics and the new quantum theory* (L. W. Codd, Trans.). London: Constable and Company Ltd.

Jordan, P. (1928). Der Charackter der Quantanphysik. *Die Naturwissenschaften [The Science of Nature]*, *16*(41), 765–772.

Josephson, P. R. (1991). *Physics and politics in revolutionary Russia*. Berkeley: University of California Press.

Langevin, P. (1923). *La physique depuis vingt ans*. Paris: G. Doin.

Lenin, V. (1976). *Collected works* (Vol. 38). Moscow: Progress Publishers.

Lenin, V. (1977). *Collected works* (Vol. 14). Moscow: Progress Publishers.

Mehra, J., & Rechenberg, H. (2000). *The historical development of quantum theory* (Vol. 6, Part 1). Heidelberg: Springer.

Planck, M. (1914). *Dynamische und Statistische Gesetzmäßigkeit*. Leipzig: J. A. Barth.

Plekhanov, G. V. (1976). *Selected philosophical works II*. Moscow: Progress Publishers.

Schrödinger, E. (1935). *Science and the human temperament*. London: George Allen and Unwin.

Schrödinger, E. (1984). *Sitzungsberichte der Preussischen Akademie der Wissenschaften, Gesammelte Abhandlungen = Collected papers, 3*. Wien: Verlag der Österreichischen Akademie der Wissenschaften, F. Vieweg.

Turnbull, H. W. (Ed.). (1960). *The correspondence of Isaac Newton II*. New York: Cambridge University Press.

Chapter 12
Materialist Dialectics and Modern Physics. Report Abstracts. The First All-Union Congress of Physicists, Odessa, August 19, 1930

Boris Hessen

1. Lenin wrote in 1908, "The essence of the crisis in modern physics consists in the break-down of the old laws and basic principles, in the rejection of an objective reality existing outside the mind, that is, in the replacement of materialism by idealism and agnosticism."[1] Lenin's analysis of the advancement of physics and his prognosis for its further development is fully confirmed by the current situation in physics.
2. Theoretical natural science cannot do without general concepts and categories. The very process of the development of natural science, i.e. the accumulation of new material leads to the necessary transformation of the old system of concepts. The material does not fit within the framework of the old worldview.
3. This conflict between new content and old forms of categories can be expressed in two ways: either in the defending of naked empiricism, in renouncing the necessity for any philosophical generalisation and in the slogan "physics is in itself philosophy," or in acceptance of all categories being conventional, without [. . .][2] objective reality, i.e. in sliding back to idealism and agnosticism.

[1] TN: *Materialism and Empirio-criticism*, Lenin (1977), p. 258.
[2] K: One word illegible.

TN: Original from Archive of the Russian Academy of Science. Ph. 1515, Op. 2. D.17. L. 1-8. Published by S. N. Korsakov, Korsakov (2019). Reproduced here with permission. Korsakov's footnotes are denoted by K or K/BH if origin unknown.

B. Hessen (✉)
Abingdon, UK
e-mail: chris.talbot@protonmail.com

C. Talbot and O. Pattison (eds.), *Boris Hessen: Physics and Philosophy in the Soviet Union, 1927–1931*, History of Physics,
https://doi.org/10.1007/978-3-030-70045-4_12

4. Mechanistic materialism was the worldview of classical physics. Even in the 70s and 80s of the last century, i.e. at the time of the domination of the mechanistic worldview, Engels, with exhaustive clarity, exposed its limitations and its insufficiency as a methodological foundation for natural science. On the basis of the analysis of physics in the beginning of the twentieth century, Lenin demonstrated that the crisis of the fundamentals of modern physics was caused precisely by the limitations and the insufficiency of mechanistic materialism. It was necessary to substitute dialectical for mechanistic materialism.
5. The vast majority of natural scientists remained unaware of works by Engels and Lenin. The crisis of mechanistic materialism remains for natural scientists a crisis of materialism. Meanwhile the only way out of the crisis is the substitution of mechanistic materialism by the higher form of materialism, i.e. dialectical materialism.
6. The long domination of mechanistic materialism and the subsequent achievements of physics in the eighteenth and nineteenth centuries convinced scientists of the classical period of natural science that this worldview provides the explanation of phenomena. This worldview was turning into an absolute truth. (W. Thomson)[3]
7. The main feature of dialectical materialism is the historical approach to all phenomena. Everything should be considered in interconnection and development. Science and scientific truth are not something absolute and given once and for all. "Science is the creative advancement of knowledge".[4] The transformation of scientific truths is a necessary stage in the process of the unlimited approximation towards absolute truth that reflects the change of socio-economic forms.
8. Mechanical materialism developed on the basis of terrestrial and celestial mechanics, i.e. on the basis of the physics of macroworld. This conditioned its limitations. The mechanical measure of things became universal. Mechanical laws became considered to be universal laws. The task of natural science, instead of the knowledge of the universal connection, unity of development and specificity of forms of motion of matter, was proclaimed to be the reduction of all phenomena, without exception, to the mechanical motion and positions of elementary particles.
9. The progressiveness of the mechanical worldview which succeeded scholastic physics was in the aspiration to establish the unity of all phenomena. However, this unity established by the mechanical worldview was purely superficial because it was achieved by the universality of a particular form of motion, i.e. mechanical motion.
10. This is why mechanical materialism was unable to resolve the problem of the real dialectical unity of the forms of motion that includes the specificity of each form and does not dissolve it in the formal unity of the reduction of all forms of motion to the mechanical one.

---

[3]K/BH: William Thomson, First Baron Kelvin, 1824–1907, a British physicist, the President of London Royal Society in 1890–1895.

[4]TN: Reference not found.

11. Mechanical materialism is unable to resolve the problem of development because it understands every form of motion as a simple sum of the mechanical motion of elementary particles.
12. Mechanical materialism was the worldview of the rising bourgeoisie which used natural science to serve the development of productive forces in order to overcome feudalism. The struggle between mechanical materialism and medieval scholasticism ideologically reflected the struggle between capitalism and feudalism.
13. As "at a certain stage of development, the material productive forces of society come into conflict with the existing relations of production . . . within the framework of which they have operated hitherto. From forms of development of the productive forces these relations turn into their fetters."[5], similarly the mechanical worldview became the fetter on the advancement of natural science.
14. This does not mean that mechanical materialism should be thrown away and pronounced absolutely wrong. It should be not thrown away but dialectically overcome by a higher level of materialism, i.e. materialistic dialectics.
15. Materialistic dialectics is a theory of development. It originates from a concept of motion that is completely different from mechanistic materialism. Dialectical materialism understands motion as a general change or development. "Motion in the most general sense, conceived as the mode of existence, the inherent attribute, of matter, comprehends all changes and processes occurring in the universe, from mere change of place right up to thinking."[6]
16. There are two conceptions of development and motion: either motion and development as a "struggle of opposites", i.e. acknowledgement (discovery) of the opposing *mutually exclusive* trends in *all* phenomena and processes of nature. "The condition for the knowledge of all processes of the world in their self-movement, in their spontaneous development, in their real life, is the knowledge of them as a unity of opposites."[7] Or ". . . development as a decrease and increase, as repetition."[8] The first concept of development is materialistic dialectics, and the second is the concept of mechanistic materialism.
17. If we analyse the historical process of the advancement of physics in the field of four cardinal problems: 1. Space, time and matter; 2. Matter and motion; 3. Structure of matter; 4. Statistical and dynamical laws; then we shall see that the entire history of the development of these four problems fully confirms the dialectical concept of the motion of matter characterised by Lenin above.

[5]TN: K. Marx, *Preface to A Contribution to the Critique of Political Economy*, 1859, Marx and Engels (1966), pp. 503–4.

[6]TN: F. Engels, *Dialectics of Nature*, Engels (1988), p. 362.

[7]TN: *Philosophical Notebooks*, Lenin (1976), p. 358.

[8]TN: ibid.

18. With regard to the problem of space, time and matter, classical mechanical physics held the view of absolute empty space containing matter as if in a box and existing separately from empty time that flowed independently from any processes. Modern physics arrives at the concept of the unity of space, time and matter.

    Space and time are the fundamental conditions for the existence of matter. Space and time are united in real synthesis in matter. Only abstract space and time exist without matter.

    Matter is not in space and time as if inside an empty box but interacts with them, thereby conditioning the structure of the space-time continuum. Thus, Kantian ideas about the a priori of Euclidean geometry are refuted and the concepts of space, time and matter first expressed by Hegel and materialistically developed by Engels are confirmed.
19. With regard to the problem of matter and motion, modern physics arrives at the viewpoint of the inseparability of motion from matter and matter from motion. Motion is a form of existence of matter. The advancement of quantum mechanics, the necessity of the introduction of zero energy and the uncertainty principle necessarily lead to the realisation that real matter is always in motion.

    Classical physics considered motion not as inherent in matter but as a result of an external force. The concept of a force was one of the main and at the same time, of the most obscure concepts of classical physics.

    Modern physics and the general theory of relativity accepts the point of view of the inseparability of motion from matter in the sense that motion is considered not to be borne by an external force but inherent in matter. Although this concept of relativity theory is relevant only to gravitational forces, the further advancement of this theory demonstrates that the evolution of the understanding of electromagnetic fields follows the same direction.
20. It is worth noting that the essence of the geodesic line principle substituted for Newtonian gravitational forces is not only that the bodies follow the geodesic lines due to their inherent motion (this might not happen in a universal field theory), but mainly that the form of the body's motion results from a complex interaction with the entire material space-time continuum.
21. With regard to the problem of the structure of matter, there is a fundamental departure from the point of view of classical physics.

    Discontinuity and continuity, as well as particle and field stood side by side in classical physics. However, the continuum had a priority. Attempts to "explain" a discrete particle, if at all, were made by making it to be a formation in the continuum. Discreteness was reduced to continuity (e.g. W. Thomson's vortex atom).

    The advancement of atomic physics brought about yet sharper contradictions (termed "dualism" by physicists) between fields and discrete particles, i.e. waves and corpuscles. Instead of deriving a corpuscle from a wave (first thought of by Schrödinger) modern physics more and more decisively adopts the view that wave and corpuscle are real opposites of the objective structure of matter, i.e. "are rooted in the very nature of an object." (J. J. Thomson)[9]

"If quantum mechanics has any peculiarity, it is that it does not decide between two modes of presentation (corpuscles and waves) but after the seeming victory of one, reinstates the other and combines both in higher unity."[10] However, it should be noted that relativity theory continues to remain with the standpoint of an ideal continuum contrary to the views that accept the objectivity of the contradictory structure of matter related above. This is the non-dialectical aspect of relativity theory. Even Einstein has to agree that the contradiction between discontinuity and continuity was never so pronounced as it is now. However, he does not speak in support of the objectivity of this contradiction.

22. With regard to the problem of physical laws, classical physics followed the view of the absolute priority of dynamical laws. Statistical laws were pronounced second-rate knowledge. A physical law was considered completely scientifically formulated when it was expressed in the form of a dynamical law.

    This traditional concept whose origins go back to Laplace and Newton is inextricably linked to the domination of a mechanical worldview. Since the motion of a discrete particle (or motion of an ideal liquid) is recognised as the main form of motion, then a dynamical law (i.e. a physical and mathematical expression of abstract mechanical determinism) also becomes the prototype of a physical law.
23. The latest development of quantum physics demonstrated the insufficiency of mechanical determinism for the expression of complex laws of the microworld. Statistical laws become a more dominant form of expression for laws of internal atomistic phenomena.
24. While classical physics saw its task in the reduction of all laws to dynamical ones, modern physics goes to the other extreme and considers the insufficiency of dynamical laws to be a collapse of determinism as a whole.
25. The dialectical view of the relationship between statistical and dynamical laws is that a real law is a synthesis of the two. Dynamical laws are laws of motion of a simple isolated body. Statistical laws are laws of motion of the aggregate as a whole. The whole consists of parts and is a product of development. Therefore, dynamical laws for the constituent parts of the aggregate form the basis for a statistical law. However, just as the whole is not reducible to a simple sum of its parts but is their synthesis, a statistical law is not reducible to a simple sum of dynamical laws but is a specific law for the whole. A law of motion of each part is irrelevant or accidental in relation to this specific law.

[9]K/BH: Joseph John Thomson, (1856–1940), a British physicist, the President of the London Royal Society in 1915–1920, a Nobel Prize winner. TN: This may refer to Thomson's "ether mass" theory—see Chap. 3, p. 40, n. 37.

[10]TN: Born (1929), p. 118. The English translation given here is in Born (1968), p. 35.

26. The dialectical concept of chance and necessity allows us to give a real synthesis of dynamical and statistical laws, and at the same time to stay on the ground of strict determinism. The latter is the cornerstone of the materialistic worldview without which the scientific knowledge of nature is impossible.
27. According to Lenin, "In brief, dialectics can be defined as the doctrine of the unity of opposites. This embodies the essence of dialectics . . ."[11]

    We see that the entire internal evolution of physics leads to the problem of the unity of opposites.

    Therefore, it is not accidental that the problems of discontinuity—continuity (wave-particle) and of statistical and dynamical laws are the core problems of modern physics.

    The study of the real processes of the motion of matter and of human knowledge leads us to the conviction that dialectical contradiction is an objective contradiction of any motion; and that "Dialectics is the science about general laws of motion of both external world and of human society." (Marx)[12]

References

Born, M. (1929). Über den Sinn der Physikalischen Theorien. *Die Naturwissenschaften (The Science of Nature)*, *17*(7), 109–118.

Born, M. (1968). *Physics in my generation*. Heidelberg: Springer.

Engels, F. (1988). *Marx engels collected works* (Vol. 25). Moscow: Progress Publishers.

Korsakov, S. N. (2019). Materialist dialectics and modern physics. report abstracts. the first all-union congress of physicists, odessa, august 19. 1930. *Epistemology and Philosophy of Science*, *56*(1), 209–215.

Lenin, V. (1976). *Collected works* (Vol. 38). Moscow: Progress Publishers.

Lenin, V. (1977). *Collected works* (Vol. 14). Moscow: Progress Publishers.

Marx, K., & Engels, F. (1966). *Selected works* (Vol. 1). Moscow: Progress Publishers.

[11] TN: *Philosophical Notebooks*, Lenin (1976), p. 222.

[12] TN: The reference to Marx is apparently mistaken, but in *Anti-Duhring*, Engels (1988), p .131, Engels writes: "Dialectics, however, is nothing more than the science of the general laws of motion and development of nature, human society and thought.".

Appendix
Biographical Notes

Bursian, Viktor Robertovich (1886–1945), a Russian and Soviet theoretical physicist. He worked on a range of physical problems, firstly using classical, then later quantum physics from 1918 to 1932 under Joffe at the Physico-Technical Institute in Leningrad (LFTI). From 1932 until his arrest in 1936 "for participation in a fascist-Trotskyite-Zinovievite organization" he was professor then director of the Scientific-Research Institute of Physics at Leningrad University. He carried out work in mineral prospecting in the 1920s, one of the founders of the technique of electrical geo-exploration. For more details see Bursian (1988). Bursian was sentenced in 1937 by the Supreme Military Court to 10 years in a labor camp, which he spent in the technical special office of the NKVD carrying out thermal calculations.

Egorshin, Vasilii Petrovich (1898–1985), born into a peasant family, joined the Russian Social Democratic Labor Party in 1915. After the revolution in 1921 he taught courses at Moscow University and from 1924 taught physics at the Communist University. Like Hessen he then studied at the IKP. He also joined the Deborin group, but turned against them in the late 1920s.

Fock, Vladimir Aleksandrovich (1898–1974), a major Soviet theoretical physicist, known internationally for his foundational work in quantum mechanics and QED, where he introduced key mathematical concepts such as Fock space. Graduating from Petrograd University where he was a postgraduate, becoming a professor there in 1932. He collaborated with the Physico-Technical Institute in Leningrad (LFTI) from 1924 to 1936 and had periods of collaboration with the Vavilov State Optical Institute in Petrograd (now St. Petersburg) and with the Lebedev Physics Institute in Moscow. He wrote the first Soviet textbook on quantum mechanics in 1931. Like Joffe, Tamm, Frenkel and all the main physicists in the Soviet Union in the 1920s and 30s, Fock did not question the standard "Copenhagen" interpretation of quantum mechanics. In the purges of 1936–7 Fock was briefly arrested but refused to condemn other physicists—for more details see Josephson (1991), pp. 312–314. By the late 1930s

C. Talbot and O. Pattison (eds.), *Boris Hessen: Physics and Philosophy in the Soviet Union, 1927–1931*, History of Physics,
https://doi.org/10.1007/978-3-030-70045-4

Fock advanced a dialectical materialist philosophical position on quantum theory, claiming that it would support Niels Bohr's approach, from which the "positivistic coating" could be removed. He even travelled to visit Bohr in Copenhagen in 1957 and came away convinced that he agreed with Bohr, after the latter's "correction of his formulations." See Graham (1971), pp. 69–110.

Frenkel, Yakov Il'ich (1894–1952) was a brilliant Soviet physicist, working at the Physico-Technical Institute in Leningrad (LFTI) from 1921 until his death. He taught himself theoretical physics and mathematics while at high school in Tsarist Russia. As a student at St Petersburg university he produced work of such calibre that it was noticed by Joffe. From 1922 until his death he published a book virtually every year, many becoming textbooks, with three translated into English. Traveling in England, France and Germany in 1925, spending a year in the United States in 1930, he became widely known internationally, especially for his work in condensed matter physics. For more details on Frenkel see Kojevnikov (2002), pp. 47–69. Frenkel leaned towards a positivist philosophy of quantum mechanics and openly criticised dialectical materialism in a lecture in 1931, for which he was repeatedly denounced by Soviet philosophers until his death.

Joffe, Abram Fedorovich (1880–1960), leading Russian and Soviet physicist. He was responsible for the establishment of the Physico-Technical Institute in Leningrad (LFTI) in 1918 and became its director. His work on the mechanical and electrical properties of crystals gave him a world-wide reputation, but he also worked on many other areas of physics and helped establish a number of research laboratories. He was prominent in the defence of physics against Stalinist attacks in the 1930s. Much of (Josephson (1991)) is concerned with the development of the LFTI, including Joffe's role.

Kasterin, Nikolai Petrovich (1869–1947), a prominent Russian and Soviet physicist in the 1920s, a proponent of mechanistic views, attempting to develop an alternative to relativity based on classical physics. He authored a book on aerodynamics and electrodynamics that was translated into English in 1937. Though opposed to Marxism he kept out of philosophical debates. Remarkably he joined with Timiryazev in the later 1930s in attacking relativity, presumably noting that Timiryazev had personal support from Stalin even though he was increasingly isolated among physicists.

Lazarev, Petr Petrovich (1978–1942), a Russian and Soviet physicist, biophysicist and geophysicist, member of the Soviet Academy of Science from 1917, first chief editor of *Uspekhi Fizicheskikh Nauk* (*Advances in Physical Sciences*) journal. When Lenin was shot and wounded in 1918 he was X-rayed in Lazarev's laboratory, one of the few still functioning. In 1931 Lazarev was arrested and interrogated because he was "not loyal enough" to the Soviet regime. Under the strain his wife committed suicide. Although Lazarev was released after six months he was removed as director of the Institute of Physics and Biophysics which he had founded, exiled for a while, then returned to Moscow in a low grade post which he held for six years. In 1938 his work was investigated by a special commission, and he remained in disgrace for the final 4 years of his life. See Ivanitskii (2009).

Lebedinsky, Vladimir Konstantinovich (1868–1937), a Russian and Soviet physicist and radio-engineer, first chairman of the Russian Radio-Engineering Society (1918), co-editor of the first Soviet Encyclopaedia (from 1926), best known for work on electrical, magnetic and radio physics.

Mach, Ernst Waldfried Josef Wenzel (1838–1916) was an Austrian physicist and philosopher, noted for his contributions to physics such as the study of shock waves. Mach developed a positivist philosophy which was very influential at the end of the nineteenth and the beginning of the twentieth century, a major influence on logical positivism and American pragmatism. His philosophy put forward the view that scientific laws merely correlated sensory perceptions, opposing the claim that such perceptions reflect actual relationships among real things, which exist independently of observation. (See Holton (1993), Chap. 1, for more on Mach's influence.) His ideas were taken up, in a form called Empirio-criticism, by some of the leaders of the Bolshevik Party, the best known being Alexander Bogdanov. They were strongly opposed by Lenin who wrote his book *Materialism and Empirio-criticism* against their "Machist" views, published in 1909.[1]

Mandelstam, Leonid Isaakovich (1879–1944) was from a Belarusian-Jewish background. His main interest was in optics and quantum mechanics. He is regarded as one of the founding fathers of Soviet physics, based at Moscow university. In 1928 he and G. Landsberg discovered the effect of the combinatorial scattering of light. Two Indian scientists C. V. Raman and K. S. Krishnan discovered the same effect, one week later than Mandelstam and Landsberg. They observed the effect in solids, liquids, and vapors, thus proving the universal nature of the effect whereas Mandelstam and Landsberg had only observed it in crystals. The Nobel Prize committee judged that Mandelstam and Landsberg were unable to provide an independent, complete interpretation for the discovery and had failed to cite the work of Raman and Krishnan so only the latter won the Nobel prize. The effect is now usually known as Raman scattering (although it is still referred to as "combinatorial scattering of light" in Russia) and is widely used in physics and chemistry.

Maksimov, Alexander Aleksandrovich (1891–1976) graduated in physics from the University of Kazan in 1916. He joined the Bolsheviks just after the October Revolution, and served as a soldier in the Red Army and as a provincial educational official. After demobilisation he obtained a post in the Commissariat of Education's division of secondary education in Moscow. He became chairman of Moscow University's Department of the History and Philosophy of Natural Science, which grew out of the informal study circle that he and Timiryazev organized in 1923. He joined the Deborin group of philosophers but began criticising them in 1928. He became a key

[1] Lenin (1977). For recent interesting research on this topic see Bakhurst (2018); Pavlov (2017).

opponent of Hessen and other leading physicists, accusing them of "idealism" and siding with pro-Stalin "Bolsheviser" philosophers after 1930.

Miller, Dayton Clarence (1866–1941) was an American experimental physicist who opposed Einstein's Theory of Special Relativity. Timiryazev argued that Miller's results validated his own criticisms of Einstein. Dayton Miller claimed that space was absolute and filled with an ether so that the speed of light would not be the same in all directions as it must be for Einstein to be correct. The earth moving through the ether would create an "ether drag", altering the speed of light in the direction of the earth's motion. Dayton Miller carried out more and more accurate experiments from the beginning of the twentieth century for more than 30 years, claiming he had discovered a tiny deviation in the speed of light. Many physicists, including Joffe apparently, were critical of his techniques and several carried out measurements that verified the original Michelson-Morley null result. Recent work suggests that the statistical techniques needed to analyse such a large amount of data collected by Miller were not available in that period. Current techniques confirm that his supposed variation in the speed of light was not statistically significant. See Roberts, T. J.,"An Explanation of Dayton Miller's Anomalous 'Ether Drift' Result", (2006). Available at https://arxiv.org/abs/physics/0608238, cited 26.02.2020.

Millikan, Robert Andrews (1868–1953), famous American experimental physicist. He received the Nobel prize in 1923 for the measurement of the charge on an electron and for his work on the photoelectric effect. There is no record that he carried out Dayton Miller type experiments suggested in Chap. 2, p. 22, so presumably the authors were mistaken.

von Mises, Richard (1883–1953) an Austrian scientist and mathematician, well known for his work in areas such as aerodynamics, hydrodynamics, probability and statistics. He held positivist views in philosophy in the tradition of Ernst Mach, closely associated with Philipp Frank and the so-called Vienna circle. In the 1920s von Mises' approach was central to the development of statistical physics and Hessen gave much energy to promoting his work, though, of course, opposing his philosophy (see Chap. 10). Richard von Mises's even more famous brother was Ludwig, an economist of the Austrian school.

Rosing, Boris L'vovich (sometimes spelt Rozing) (1869–1933) graduated from the department of physics and mathematics of the University of St. Petersburg in 1891. He taught at the St. Petersburg Institute of Technology (1894–1918 and 1924–31) and a number of other higher educational institutions. From 1897 he began experimenting with the electrical transmission of images over a distance and was an early pioneer of the television. In 1911 he successfully built a simple television using a cathode ray tube, one of the earliest to do so. He continued his research until 1931 when he was exiled to Archangel, accused of being a counter-revolutionary. He died in exile. The content of his talk (Chap. 2, p. 22) is not known, but no doubt would be very interesting. Advanced and retarded potentials arise in the theory of electromagnetism and can be shown mathematically to generate electromagnetic fields. They are widely studied in theoretical physics. The physical interpretation of advanced potentials is problematic as it implies an influence in the present coming from future times.

Schukarev, Alexandr Nikolaevich (1864–1936), a Russian/Ukrainian physical chemist. He graduated from Moscow University in 1889 and worked there from 1891 to 1909. Beginning in 1911, he was a professor at the Kharkov Technological Institute (now the Kharkov Polytechnic Institute). Apart from research in physical chemistry he is known for developing, prior to World War I, the "logical thinking machine"—an improved version of William Stanley Jevons's logic piano. For more details see Shilov and Silantiev (2014). According to Shilov and Silantiev, Schukarev was criticised in the mid 1920s as an opponent of dialectical materialism for the view—which he did not hold–that thinking machines could replace humans. Under the pressure for "Marxist" orthodoxy, Schukarev, who was something of an iconoclast, increasingly upset the authorities and was forced to retire in 1931.

Semkovskii, Sergei Yurievich (1882–1937), was the leading Marxist philosopher in the Ukraine. Semkovskii, real name Bronstein, was a cousin of Leon Trotsky. He wrote the first book taking a dialectical materialist approach to relativity in 1924, on which Hessen is said to have based his work. Semkovskii was particularly concerned with winning natural scientists to Marxism. A Mechanist in the mid-1920s he switched over to the Deborinites in 1929. Semkovskii was arrested on March 3 1936, accused of "Menshevizing idealism" and membership of a Trotskyist terrorist organisation. Shot on March 9 or 18, 1937.

Shaposhnikov, Konstantin Nikolaevich (1880–1957) was a Russian physicist prominent in optical physics, member of the Moscow Mathematical Society, and a proponent of the "mechanical theory of light quanta".

Skvortsov-Stepanov, Ivan Ivanovitch (1870–1928) joined the Bolshevik Party in 1904, and was arrested and exiled many times. He worked with Bazarov (a supporter of Empirio-criticism, and which Stepanov definitely opposed) in translating the three volumes of Marx's *Capital* into Russian. He was known as a propagandist rather than a theoretician, a populariser of Marxist economics. In the famine period of 1922, when the Soviet economy was in a state of collapse, he wrote *The Electrification of the R.S.F.S.R. in Connection with the Transitional Phase of the World Economy*, a book which Joravsky states "contributed to Bolshevik self-confidence" with an enthusiastic preface written by Lenin. It set out a vision for Soviet science and technology. In the 1920s, until his death in 1928, Stepanov was particularly concerned with the relation between Marxism and the natural sciences, as well as writing popular material against religion. He was regarded as a leader of the Mechanists.

Strum, Lev Yakovlevich (1890–1936), a Ukrainian theoretical physicist and philosopher, Head of the Theoretical Physics Department of Kiev University, who carried out research in relativity theory, quantum, atomic, and nuclear physics, thermodynamics, and physics methodology. In 1925–1926, L. Ya Strum made contact with the philosopher Sergei Yurievich Semkovskii (see above). In March, 1936, one of the key show trials instigated by Stalin was against an alleged "counter-revolutionary Trotskyist terrorist organization in Ukraine". Semkovskii was arrested first, followed shortly by Strum. Strum was charged with being "an active member of the counter-revolutionary Trotskyist-Menshevist underground networks in Kiev and directly communicating with the underground center," and for distributing Trotskyist propaganda. The charge was broadened to include allegedly working with the Gestapo and carrying out the assasination of Kirov. Strum was executed in October 1936, having been forced to plead guilty. This information is contained in Malykin et al. (2012). The article explains that Strum was the first to consider the hypotheses of superluminal (faster than light) velocities. (Apparently the generally accepted idea amongst physicists is that they were first discussed in the 1960s). Strum made proposals showing how to interpret them without invalidating Einstein's Theory of Special Relativity and the principles of causality. Particles moving faster than light are called tachyons and so far have never been discovered in reality.

Tamm, Igor Yevgenyevich (1895–1971) was a prominent Soviet physicist, joint winner of the Nobel Prize for the discovery of Cherenkov radiation in 1934 (the radiation that results when a charged particle, such as an electron, travels through a dielectric—an electrical insulator that can be polarized by an applied electric field—with a speed greater than that at which light propagates in the dielectric). A supporter of the 1917 revolution, after graduation Tamm began as a physics professor at the Second Moscow State University in 1923. From 1934 until his death he was the head of theoretical physics at the Lebedev Physics Institute in Moscow. Apart from Cherenkov radiation he made a number of important contributions in relativity and quantum physics, and opposed the various attacks that were made on physics in the 1930s.

Tartakovsky, Petr Savvich (1895–1940), a Russian physicist, author of "Quanta of light" and several other monographs. He worked in Leningrad Physico-Technical Institute until 1929, then Joffe offered him a post at Tomsk in Siberia where he headed a team experimenting on the quantum properties of electrons. In 1937 he returned to Leningrad where he died in 1940. See Vaisburd, D.I., "Academician S.P. Bugaev", *Tomsk Polytechnic University, TPU and Scientific Achievements online newsletter* (2000): http://www.lib.tpu.ru/fulltext/v/Tomsk_polytechnic/2000/N6a19_full.pdf cited 26.02.2020.

Timiryazev, Arkady Klimentyevich (1880–1955), by the mid 1920s one of the leaders of the Mechanist faction of Soviet philosophers who opposed the Deborin group. He joined the Communist Party in 1921 and became professor of physics at Moscow State University. The Timiryazev Institute, named after his father Kliment Arkadievich Timiryazev (1843–1920) an internationally renowned Darwinian biologist and supporter of the Russian Revolution, became a centre for Mechanist views. Timiryazev took an extreme position. He claimed to support dielectical materialism and considered that the key developments in twentieth century physics, relativity and quantum theory, were false and "idealist". He targeted Einstein especially as an idealist, pointing out that he had supported Mach's philosophy at the beginning of the century. Timiryazev embraced classical physics as the only valid materialist standpoint, including Newtonian absolute space and time, claiming that it could explain all the results of modern physics.

Vavilov, Sergey Ivanovich (1891–1951) was a leading Soviet scientist from the 1930s until his death. He became a member of the USSR Academy of Sciences in 1932 and head of the Lebedev Institute of Physics in 1934. He was appointed member of the Supreme Soviet from 1946, and was a recipient of four Stalin Prizes in the 1940s and 50s. As a physicist specialising in optics in Moscow in the 1920s he was one of the first translators of Einstein and published a popular presentation, *The Experimental Foundations of the Theory of Relativity* in 1928. Though opposed to Timiryazev he did not refer to philosophical issues nor take part in such debates. He was not regarded as an outstanding scientist in contrast to his geneticist brother Nikolai Ivanovich, famous for his opposition to the charlatan Lysenko. Nikolai Ivanovich died of starvation in prison in 1943 after condemnation by Stalin. Though primarily an administrator Sergey Ivanovich was co-discoverer with Pavel Cherenkov of what is now called Cherenkov radiation in 1934, and if he had lived would have shared the Nobel prize with Cherenkov and Tamm in 1958. For a discussion comparing his career with that of his brother, see Kojevnikov (2002), Chap. 7.

References

Bakhurst, D. (2018). On Lenin's materialism and empirio-criticism. *Studies in East European Thought*, *70*(2), 107–119.

Bursian, E. V. (1988). Viktor Robertovich Bursian (1886–1945). *Soviet Physics Uspekhi*, *31*(6), 555–557.

Graham, L. R. (1971). *Science and philosophy in the Soviet Union*. London: Allen Lane.

Holton, G. (1993). *Science and anti-science*. Harvard University Press.

Ivanitskii, G. R. (2009). Ninety years of biophysics in Russia: The way it was. *Herald of the Russian Academy of Sciences*, *79*(4), 394–401.

Josephson, P. R. (1991). *Physics and politics in revolutionary Russia*. Berkeley: University of California Press.

Kojevnikov, A. B. (2002). *Stalin's great science. The times and adventures of Soviet physicists*. London: Imperial College Press.

Lenin, V. (1977). *Collected works* (Vol. 14). Moscow: Progress Publishers.

Malykin, G. B., Savchuk, V. S., & Romanets(Shcherbak), E. A. . (2012). Lev Yakovlevich Strum and the hypothesis of the existence of tachyons. *Physics-Uspekhi*, *55*(11), 1134–1139.

Pavlov, E. V. (2017). "When was Caesar born?" Theory and practice of truth in Plekhanov and Bogdanov. *Stasis*, *5*(2), 50–79.

Shilov, V. V., & Silantiev, S. A. (2014). Reasoning vs. Orthodoxy, or, The Lesson from the Fate of Russian "Reasoning Machine". In Kimppa, K., Whitehouse, D., Kuusela, T., & Phahlamohlaka, J. (Eds.) *ICT and Society*, (pp. 191–202). Berlin, Heidelberg: Springer Berlin Heidelberg.

The manufacturer's authorised representative in the EU is Springer Nature Customer Service Centre GmbH, Europaplatz 3, 69115 Heidelberg, Germany. If you have any concerns regarding our products, please contact ProductSafety@springernature.com

Printed and bound by CPI Group (UK) Ltd, Croydon, CR0 4YY
22/09/2026
02221061-0020